D1556928

Communities and knowledge production in archaeology

Manchester University Press

Social Archaeology and Material Worlds

Series editors
Joshua Pollard and Duncan Sayer

Social Archaeology and Material Worlds aims to forefront dynamic and cutting-edge social approaches to archaeology. It brings together volumes about past people, social and material relations and landscape as explored through an archaeological lens. Topics covered may include memory, performance, identity, gender, life course, communities, materiality, landscape and archaeological politics and ethnography. The temporal scope runs from prehistory to the recent past, while the series' geographical scope is global. Books in this series bring innovative, interpretive approaches to important social questions within archaeology. Interdisciplinary methods which use up-to-date science, history or both, in combination with good theoretical insight, are encouraged. The series aims to publish research monographs and well-focused edited volumes that explore dynamic and complex questions, the why, how and who of archaeological research.

Previously published

Neolithic cave burials: Agency, structure and environment
Rick Peterson

The Irish tower house: Society, economy and environment, c. 1300–1650
Victoria L. McAlister

An archaeology of lunacy: Managing madness in early nineteenth-century asylums
Katherine Fennelly

Forthcoming

Images in the making: Art, process, archaeology
Ing-Marie Back Danielsson and Andrew Meirion Jones (eds)

Early Anglo-Saxon cemeteries: Kinship, community and mortuary space
Duncan Sayer

Urban Zooarchaeology
James Morris

An archaeology of innovation: Approaching social and technological change in human society
Catherine J. Frieman

Communities and knowledge production in archaeology

Edited by Julia Roberts, Kathleen Sheppard, Ulf R. Hansson and Jonathan R. Trigg

Manchester University Press

Published by Manchester University Press
Altrincham Street, Manchester M1 7JA

www.manchesteruniversitypress.co.uk

British Library Cataloguing-in-Publication Data
A catalogue record for this book is available from the British Library

ISBN 978 1 5261 3455 4 hardback
ISBN 978 1 5261 3456 1 open access

First published 2020

Typeset in 10.5/12.5 Sabon LT Std by
Servis Filmsetting Ltd, Stockport, Cheshire
Printed in Great Britain by
TJ International Ltd, Padstow

Contents

Figures

Where, in this table, the symbol © shows that copyright has been asserted, all rights in the figure have been reserved and permission to use the figure must be obtained from the copyright holder.

Contributors

Elisabeth Arwill-Nordbladh
Elisabeth is professor emerita in archaeology at the Department of Historical Studies, University of Gothenburg, Sweden. She earned her PhD with the dissertation 'Constructions of gender in the Nordic Viking age: Past and present' (in Swedish, English summary). Her research interests focus primarily on gender studies, the historiography of archaeology, the Scandinavian Viking Age, memory and the agency of material phenomena. She has published several articles about Scandinvian archaeology's formative years from a gender-critical viewpoint. Her recent publications include 'Golden node: linking memory to time and place', in *Counterpoint: Essays in Archaeology and Heritage Studies in honor of Professor Kristian Kristiansen* (Oxford: Archaeopress, 2013).

Thea De Armond
Thea received her PhD in classical archaeology from Stanford University, California. Her research centres on histories of archaeology in central and eastern Europe, particularly its politics and geographies. She is also an excavating archaeologist, mostly working in eastern Europe and the Black Sea region. Currently, Thea is a research associate with Mapping the Grand Tour, a project to create a digital database and research tool from John Ingamells' *Dictionary of British and Irish Travellers in Italy, 1701–1800* (New Haven, CT: Yale University Press, 2003).

Laurien de Gelder
Laurien is junior curator at the Allard Pierson Museum in Amsterdam, where she works on the re-installation of the permanent collection. As

a Mediterranean archaeologist she specialises in the history of archaeology, with a focus on the history of Dutch archaeological practices in the Mediterranean basin. In collaboration with the Archaeological Service of Rome she worked as a trainee at the Royal Netherlands Institute in Rome on the Santa Prisca project (autumn 2015), which unlocked and re-examined the legacy data of the first Dutch excavations in Italy (1952–66).

Francesca de Tomasi

Francesca is an assistant exhibitions and collections manager at Musei Capitolini in Rome (Italy). In 2015 she received her PhD with a thesis on the export and trade of antiquities in post-unification Italy ('Esportazione e commercio di antichità: Roma, 1870–1909. Aspetti normativi, contesto socio-culturale e committenza straniera'). She was associate research fellow at the Italian Academy for Advanced Studies in America at Columbia University, New York in 2016 and post-doctoral fellow at the Istituto Italiano per gli Studi Storici in Naples in 2015. She has published several articles on the late-nineteenth-century antiquarian market. Her current project investigates the history of Roman antiquities collections in America and the connections between American museum culture and the flourishing late-nineteenth- and early-twentieth-century antiquarian market.

Anna Gustavsson

Anna is a PhD candidate in the Department of Historical Studies at the University of Gothenburg. She has an MA in Classical Archaeology and Ancient History, an MA in North European Archaeology, and has held a scholarship at the Swedish Institutes of Classical Studies in Rome and Athens. Her current work analyses the development of archaeology in the late-nineteenth and early-twentieth centuries and the production and mediation of knowledge, focusing on the interrelations of Swedish and Italian scholars. Her wider research interests include the history of archaeology, museums and collections, the role of archaeology in the present, Classical reception studies and Mediterranean archaeology. Anna has been engaged in archaeological fieldwork and the heritage sector since 2002, working in Sweden, Italy and Greece.

Ulf R. Hansson

Ulf is Director of the Swedish Institute of Classical Studies in Rome and a senior research fellow in Classics at the University of Texas at Austin. His fields of interest include Etruscan and Roman art and archaeology, the cultural history of antiquarianism and archaeology, the history of collecting and collections, and the early modern reception of Greek and

Roman art and culture. He has worked extensively on German classical archaeology, especially Adolf Furtwängler and the Munich school, and in 2013 he organised the major international conference Classical Archaeology in the Late Nineteenth Century (Berlin: De Gruyter, forthcoming). His current research focuses on antiquarianism and proto-archaeology in early-eighteenth-century Rome.

Vladimir V. Mihajlović

Vladimir is a research assistant at the Institute for Balkan Studies of the Serbian Academy of Sciences and Arts (SASA) in Belgrade. His research interests lie in the fields of the theory and history of archaeology, such as theoretical and methodological aspects of archival research in archaeology, the political use of the past, transfer of knowledge between Balkan academic circles and the rest of Europe, and the subsequent dissemination of that knowledge (through school textbooks in particular). Currently, his work focuses on the antiquarian/pre-disciplinary roots of knowledge of the discipline in Serbian and Balkan archaeologies.

Monika Milosavljević

Monika is an assistant professor in the Department of Archaeology, Faculty of Philosophy at the University of Belgrade as well as a research assistant on the project Archaeological Culture and Identity in the West Balkans, funded by the Ministry of Education, Science and Technological Development of the Republic of Serbia. She lectures on archaeological theory and methodology. Her research interests lie in the political usage of archaeology, the history of Serbian/Yugoslavian archaeology, sociocultural evolution, the archaeology of identity and archaeological theory in general. In recent years, she has focused on the history of ideas in Serbian and Yugoslavian archaeology. Moreover, she is interested in theory and methodology in the history of science, particularly in the work of Ludwik Fleck. She is currently researching the intersection of human–animal relations, referred to as the 'animal turn', in medieval studies.

Julia Roberts

Following on from her PhD work, Julia works as an independent scholar, researching the history of British archaeology during the first half of the twentieth century. In particular, she concentrates on colonialism, gender, class, race and access to archaeology over this period. Julia has extensively researched how ideas from contemporary society impacted on twentieth-century archaeology and influenced the interpretations of the past offered by archaeologists. She is also a part-time lecturer at the University of Central Lancashire, a freelance archaeologist specialising in editing and illustration and an administrator for the Histories of Archaeology Research Network (HARN), producing the weekly blog post.

Kathleen Sheppard
Kathleen got her PhD in the Department of History of Science at the University of Oklahoma in 2010 and currently works at the Missouri University of Science and Technology (Missouri S&T) in Rolla, Missouri, USA. Her most recent work is an edited volume of correspondence, *My dear Miss Ransom: Letters between Caroline Ransom Williams and James Henry Breasted, 1898–1935* (Archaeopress, 2018). Her first book was *A Woman's Work in Archaeology: The Life of Margaret Alice Murray* (Lexington Books, 2013), a scientific biography of the first woman lecturer in University College, London's Egyptology department. Her current scholarship continues to focus on the history of Egyptology, but she has shifted to the development of the discipline in the United States in the early twentieth century. She is a contributing editor to the online magazine *Lady Science*.

James E. Snead
James is Professor of Anthropology at California State University, Northridge, California, USA. Awarded the PhD at UCLA in 1994, he has held numerous fellowships and post-doctoral appointments, and has received funding from the National Science Foundation and the Wenner-Gren Foundation for Anthropological Research. His study of the history of archaeology in the American Southwest, *Ruins and Rivals*, was published by the University of Arizona Press in 2001 and was followed by numerous related articles and book chapters. Current research emphasises cultural landscapes, historical archaeology of the American West, and public engagement with antiquities. New publications include 'The original Jones Boys: archaeologies of race and place in 19th century America' (*World Archaeology*, forthcoming). His most recent book project, entitled *Relic Hunters: Archaeology and the Public in 19th Century America*, was published in 2018 by Oxford University Press.

Jonathan R. Trigg
Jonathan studied at Lampeter, Liverpool and Glasgow and currently works at the University of Liverpool. He has extensive professional and academic experience in both prehistoric and historic archaeology, specialising on the history of archaeology, the archaeology of conflict and British and Irish prehistory. He has published a number of articles and book chapters on these subjects, and has edited two books. He has conducted fieldwork in Britain and Ireland. He is an

administrator of the Histories of Archaeology Research Network, Deputy Director of the Scottish Episcopal Palaces Project and a Fellow of the Society of Antiquaries of Scotland. His research examines the epistemology/historiography of British prehistory, Neolithic–Bronze Age Britain and Ireland, with particular reference to the Beaker culture, and the commemoration of conflict. He is currently examining processes of commemoration relating to battlefield burials and temporary gravemarkers of the First World War. He has a project investigating the archaeology of the Great Budworth region through time. Jonathan is also involved in projects looking at the landscape of the Fetternear region of Aberdeenshire, along with a number of other sites in north-east Scotland, and is researching the life of W.J. Varley.

Arthur Weststeijn
Arthur Weststeijn is assistant professor in Italian Studies at Utrecht University in The Netherlands. Previously, he was Head of Studies in History at the Royal Netherlands Institute in Rome. He studies the relationship between politics and intellectual culture, with a specific focus on the Classical tradition. Together with Frederick Whitling, he recently published a cultural and archaeological history of the central station in Rome: *Termini. Cornerstone of Modern Rome* (Rome: Quasar, 2017).

Acknowledgements

The editors would like to thank each of the authors in this book for their hard work, patience and participation. Each of us would also like to express thanks to our editors at Manchester University Press for working with us diligently through this process. The Histories of Archaeology Research Network (HARN) is full of amazing scholars whose work is important to understanding the global networks we all study and work in regularly. Thank you for all you do as colleagues from afar, or friends and colleagues close by.

Abbreviations

ABKF	Akademiskt Bildade Kvinnors Förening (Swedish association for women with degrees).
AIAC	Associazione Internazionale di Archaeologia Classica.
ANT	actor-network theory.
ASSAR	archive of the SSBAR *(see below)*.
ATA	Riksantikvarieämbetet, Stockholm, Antikvarisk-topografiska arkivet.
AUK FF UK	Archives of Charles University, Prague, Philosophical Faculty.
BIASA	Biblioteca di Archeologia e Storia dell'Arte, Rome.
CIAAP	International congress on Anthropology and Prehistoric Archaeology.
CLE	Collected Letters on Ethnology, Record Unit 58. Smithsonian Institution Archives, Washington, DC.
CRP	Charles Rau Papers, Record Unit 7070. Smithsonian Institution Archives, Washington, DC.
DAI	Deutsches Archäologisches Institut, Rome.
F	French National Archives, Paris.
KNIR	Royal Netherlands Institute in Rome.
LKPR	Landsföreningen för Kvinnans Politiska Rösträtt, United Swedish Suffragette Organisation.
MFA	Museum of Fine Arts, Boston, Mass.
MMA	Metropolitan Museum of Art, New York City.
MÚA AV ČR	Masaryk Institute and Archives of the Academy of Sciences of the Czech Republic, Prague.

SSBAR	Soprintendenza Speciale per i Beni Archeologici di Roma.
Unione	Unione degli Istituti di Archeologia, Storia e Storia dell'Arte.
ZWO	Netherlands Organisation for Pure Scientific Research.

Introduction: clusters of knowledge

Julia Roberts and Kathleen Sheppard

This edited volume is the first to apply scientific network theories to the history of archaeology. As an innovative approach to historiography it takes its place amongst recent studies that have transformed the discipline. Using theories including those of Ludwik Fleck, David Livingstone, Michel Callon and Bruno Latour, the authors of the following chapters have taken an unprecedented approach to their subjects: rather than looking at individuals and groups biographically or institutionally, or accepting that this is simply how archaeology *was*, these studies look at how networks are formed and how this in turn impacts on how archaeological knowledge is generated and disseminated. This original perspective has yielded novel and surprising insights into the history of archaeology which, we believe, will become the foundation of a new appreciation of the complexity of archaeology's history.

Studies of the histories of archaeology have dramatically increased in recent decades. Prior to Bruce Trigger's ground-breaking *A History of Archaeological Thought* in 1989, students had few texts to consult and, of those, many were repetitive, focussing on a few key names, generally 'great men' of archaeology credited with being the 'father' of whatever archaeology they espoused. The studies had little to offer more rigorous and theoretical archaeologists, particularly those interested in gender, race or class and how those with more marginal status access archaeology. In this climate, *A History of Archaeological Thought* quickly became a seminal work, *the* go-to textbook for students, lecturers and researchers. While it is undoubtedly flawed, as any pioneer text inevitably is, *A History of Archaeological Thought* provided archaeologists with a social, economic and politically grounded intellectual history of their origins and, perhaps more importantly, it gave succeeding generations of researchers the justification to investigate archaeological history (e.g. Patterson, 1993; Díaz-Andreu and Sørensen, 1998; Schlanger and Nordbladh, 2008; Abadía, 2009; Klejn, 2012).

Trigger's work sparked a revolution in writing archaeological histories. Those of us who felt there was still more to say – different people, different methods and different ideas to be investigated – now had an authoritative foundation from which to begin our work. The 1980s and 1990s saw an explosion of interest: workshops, seminars and conference sessions were organised; while many of these – most notably the Cambridge Critical History Sessions – remain unpublished, they did give rise to several important volumes (e.g. Christenson, 1989; Kohl and Fawcett, 1995; Díaz-Andreu and Champion, 1996; Díaz-Andreu and Sørensen, 1998; Härke, 2000; Schlanger and Nordbladh, 2008) and countless journal articles (e.g. Bar-Yosef and Mazar, 1982; Arnold, 1990; Evans, 1989, 1990, 1998). These works that more specifically analyse the history of the practice of archaeology in various contexts in turn have inspired ever more sophisticated and complex readings of history. Micro-histories of Cambridge University's archaeology department (Smith, 2009) and finely drawn contextual biographies (Sheppard, 2013) investigate how interpersonal relationships impact who practises archaeology as well as their methods and theories. Examinations of fieldwork practice (Lucas, 2001), including fringe archaeology in Britain before the Second World War (Stout, 2008), broaden the picture of how archaeology was performed in the field. Moreover, there has been a move away from the assumption that 'archaeology' means solely European prehistory (Hall, 2000; Mizoguchi, 2011). Histories of historical archaeology are appearing and the history of Classical archaeology has been gathering steady momentum (Schnapp, 1996, 2002; Gran-Aymerich, 1998, 2001, 2007; Orser, 2004; Dyson, 2004, 2006; Hicks and Beaudry, 2006), ensuring that historians and practitioners of archaeology get a more well-rounded view of the whole field, as opposed to a snapshot at a distinct period. Egyptology has, unsurprisingly, proved to be a productive area of enquiry with a very particular history, one which has enormous public appeal in the form of both broader histories (e.g. Thompson, 2015) and more specific explorations into particular episodes of colonialism, education, field practice, biography and mummy studies (MacDonald and Rice, 2003; Ucko and Champion, 2003; Day, 2006; Murray and Evans, 2008; Carruthers, 2014; Murray, 2014). However, this heroic thread of disciplinary history has been contested by imperialists and nationalists alike (Mitchell, 1991; Reid, 2002, 2015; Jeffreys, 2003) and has the additional complexity of having largely been the work of foreign investigators relying on a native workforce (Drower,

1985; James, 1992; Thompson, 2008; Quirke, 2010; Abt, 2011; Adams, 2013). From the start, authors of these histories have been, largely but not exclusively, archaeologists who wished to explore the history of their own discipline as performed in the field. In doing so, they laid the historiographical groundwork.

Inevitably, as more nuanced and informed histories of archaeology have been written, there has been increased interest in the subject both from within and outside the discipline. External historiographers have brought different methods and theories to the writing of these stories. There is a wider range of critical approaches and analytical frameworks available to bring about new angles of inspection, such as feminist history, queer theory, science and technology studies, and political history. In turn, these new approaches have been adopted and utilised by archaeologists who now define themselves as historians of archaeology rather than simply archaeologists interested in the discipline's history. These alternative perspectives utilise new research agendas, theoretical foundations and critical approaches that incorporate archaeological practice into the narratives of political, economic, social and cultural history (Patterson, 1995; Schmidt and Patterson, 1995; Meskell, 1998; Roberts, 2005), histories of education (Janssen, 1992), histories of the professionalisation of science (Levine, 1986), *conversaziones* (Alberti, 2003), feminist and gender theory (Gero and Conkey, 1991; Cohen and Joukowsky, 2004) and more. In doing so, it becomes clear that archaeology's history is not a simple, teleological tale of heroic excavators digging up remnants of a past civilisation, but instead contains an exciting, multi-disciplinary and multi-faceted complexity of stories, which have opened up the history of archaeology and revealed so much more about our past.

In recent years archives have become a focus of critical histories with archaeologists debating both what constitutes an archive and how it should be utilised (Schlanger and Nordbladh, 2008; Lucas, 2012; see also Derrida and Prenowitz, 1995; Ketelaar, 2001; Manoff, 2004). There are practical and chronological histories of institutions, societies, museums, fieldwork, archaeological theories and archaeological sites. In recent years, more theoretical histories have been written, drawing on processualism, post-processualism, personhood theory, historiography, and the philosophy and sociology of science, all of which help historians analyse the formation of these fields in new and innovative ways. Many of these works inspect the acquisition of artefact collections and how those collections shaped knowledge of particular cultures or the practice of other sciences (Shepherd, 2002, 2003; Moser, 2006; Alberti, 2009; Challis, 2013; Stevenson, 2019). In fact, the angles from which to view the history of the discipline of archaeology have become so numerous that, as Hamilakis has said, 'there is not one but many histories of archaeology' (2010: 893).

This volume takes its place amongst these studies introducing alternative readings of archaeological historiography, but it does so in an entirely innovative way. We present individual case studies that collectively analyse the process of *how* archaeological knowledge is generated based on where and by whom it is created. Each of the chapters, and therefore the entire volume, uses as its theoretical foundation the history and philosophy of science, in which there is a rich tradition of investigating the role of communication among practitioners using Ludwik Fleck's theory of 'thought collectives', Bruno Latour's actor-network theory, and the geography of knowledge (Fleck, 1979 [1935]; Livingstone, 2003; Latour, 2005; Shapin, 2010). Fleck (1979 [1935]) argued that the production of scientific knowledge is largely a social process which depends upon not only the actors themselves, but the cultural and historical contexts of their work. Related to this, Latour's actor-network theory argues that these interactions between and among scientific practitioners shifts and changes depending on which actors are present in a given context, thus making up the network of people at a given place or time. Actor-network theory is careful not to explain the *how* or the *why* of network formation or behaviour, but it allows scholars to interrogate the relationships within networks simply by providing the *who* and the *where*. If we may define these fields of study by what questions they answer, new knowledge would clearly answer the question where knowledge is created, as well as the questions who gets to participate in which investigations, and why and how they are able to take part.

Thought collectives, actor networks, and studies of place are crucial to the foundation of the social studies of scientific networks – a complex theoretical framework used to analyse sociological phenomena in a historical context. Although historians have been studying the history of archaeology using some of these ideas for almost a decade, this volume of collected works is the first of its kind in the field to use these theories as a unified web to tell the stories of some familiar practitioners, sites and institutions. However, it makes no claim to be comprehensive. We understand that there are limits to the chapters here, especially geographically speaking, and we understand that we have not incorporated all of the network and practice theories presented in this introductory essay into the chapters that follow. Instead, all the authors of this volume aim to use the examples in the chapters presented here to continue on a larger, more collective, scale an important conversation about practice that needs to take place in archaeology. In order to answer the historical questions of participation, of knowledge formation, of the importance of place in the development of the discipline and of the centrality of historical context in the story of archaeology in the past, we must use different theoretical tools.

Each of the chapters included in this publication presents a vignette of a network in which knowledge is exchanged and the effects these networks have on other groups and single actors. Networks create, share, consider and work through knowledge systems. Martin J. Rudwick's *The Great Devonian Controversy* (1985) details the 'shaping of scientific knowledge among gentlemanly specialists' in and around Devonshire between the early and the middle nineteenth century. Using these groups of gentleman scientists, Rudwick analyses how networks operate and behave when creating knowledge. He argues that these short-lived networks have a long-lasting impact on scientific thought because of their presence in time and space.

David Livingstone's and Charles Withers' work on the geography of knowledge (2011) expands on the idea of knowledge creation in particular places, and how people operating in those spaces are affected by locality. Where science is done depends on who is able to or allowed to participate in the creation and communication of knowledge; the reverse is also true, that is, who is allowed to create knowledge depends on where science is done (see also Livingstone, 2007). The chapters in this volume collectively use geography of knowledge to determine how relationships within scientific networks operate depending on where they were built, where they operate, and where and how their knowledge is spread. To do this, it is crucial to understand who is interacting at different types of site, such as universities, excavation sites, museum offices, private homes, hotels or formal scholarly meetings. But what happens once those ideas leave specific spaces?

Throughout the history of science, practitioners – both amateurs and professionals – have shared knowledge with their scholarly communities through various forms of interaction such as publications, conferences, seminars, lectures and exhibitions. Bernard Lightman's seminal *Victorian Science in Context* (1997) focuses on public lectures and public exhibits during the late-nineteenth century as key spaces in which scientists engaged with particular audiences. These popular public events are particular points of analysis for other historians as well (e.g. Levine, 1986; Sweet, 2004). Other, smaller networks were clearly established by a variety of means. Rudwick (1985) clearly studies the power of in-person conversations for spreading knowledge and building professional connections. Mary Terrall examines the power of private and semi-public spaces for doing science in eighteenth-century France (2014). And Samuel J.M.M. Alberti's work focuses on various institutions – museums, semi-private *conversaziones*, academic societies

– and their roles in the spreading of new knowledge about the natural world (2007, 2009, 2017). Each of these key texts investigates not only the spreading of knowledge but also the responses it provokes, which, arguably, constitutes an open dialogue indispensable for the community's accumulation and revision of collective knowledge. However, preceding such public events information is gained and exchanged by informal clusters or networks of scholars, individuals and groups, who generate and communicate knowledge and ideas both within the system and with external actors and communities.

The creation of and activities involved in these networks and communities are central to the chapters in this volume. Studying the groups of colleagues, assistants, students and staff is not new to the history of science, but in the case of archaeology it bears some explicit discussion here. Scholars who study present-day scientific networks argue that the best way to visualise their connections is through tracing joint publication and reviews of those publications (Newman, 2001; Glänzel and Schubert, 2005). However, it is often difficult to trace more personal contact. Many further argue that correspondence is a key piece of evidence in understanding how networks interacted with each other outside publications, that is, out of the public eye. Jim Secord's *Victorian Sensation* (2000) traces the acceptance of Charles Darwin's *On the Origin of Species* after its publication in 1859. To do this, he relies heavily on Darwin's correspondence with scientists and laypeople throughout his life. Darwin was a prolific correspondent and there are thousands of letters authors have used and continue to use as important sources for studying the behaviour of networks. Another of Darwin's biographers, Janet Browne, has recently argued that studying correspondence among scientific networks allows 'the prospect of reconstructing patterns of sociability with due appreciation to the structure of the society in which they emerged' (2014: 169). More generally, in Ruth Finnegan's edited volume *Participating in the Knowledge Society* (2005), the individual chapters taken as a whole study how the knowledge society is 'engaged in active knowledge building outside the university walls' (1). The authors are concerned with how researchers interact with one another and with scientific information away from their professional offices and laboratories. That volume, much like this one, is a general one by design, dealing with studies of communities of amateurs and professionals within (or outside) universities and industry. In general, the volume studies both written and spoken conversation among these groups, and the picture it presents of these widely varied communities is one of a unified endeavour to create knowledge.

But these works do not deal with archaeology, which is a particular kind of practice. In archaeology, the main groups of scholars who influence each other tend to gather in the field, in ephemeral groups in which some members are permanent fixtures every dig season, others come and go, and still others only appear once, briefly, and then vanish into the dust of the site and archive. Their connections do not necessarily appear in joint publications, and are therefore hard to trace. But by doing archival work and reading diaries and letters, site reports and more, we gain insight into schools of thought, in order to better understand who is sharing ideas, how these are being shared and who is participating. Every chapter in this publication illustrates this expansion and diversity in the inclusion of new methods in the history of archaeology; each of them in turn concentrates on a particular aspect of archaeological history: the critical examination of modes of knowledge exchange between individuals and groups and how this affects the trajectories of their public ideas about material culture and past civilisations.

Outline of chapters

The individual chapters in this volume focus on the networks archaeologists create and how communication among them affects the work archaeologists produce. Much of the evidence used in this volume comes from archival sources rather than published ones since these exchanges take place in person or through correspondence. As a unit, the chapters argue that the informal character of these gatherings inspired the generation of ideas and thus markedly affected the process of knowledge production in other, equally significant, ways than scholarship produced within more formal contexts. Each author, nevertheless, takes a unique approach to the topic. The chapters can be roughly grouped into those that discuss institutions – by Milosavljević, Snead, de Tomasi – and those that discuss individuals – by De Armond, Gustavsson, Hansson, Arwill-Nordbladh, Sheppard, Mihajlović, Trigg. Connecting the two groups is Weststeijn and de Gelder's chapter, in which they discuss the work of two individuals, Carl Claudius van Essen and Maarten Vermaseren, via the Royal Netherlands Institute in Rome and the wider, post-Second World War political, archaeological and economic networks. The emphasis on individuals does not imply that they were more important than institutions, and it has to be admitted that the division is not always clear-cut. So, while Gustavsson, Hansson, Arwill-Nordbladh and Sheppard discuss individual scholars they do so partially within the context of institutions. Snead, Milosavljević and de Tomasi focus on institutions but refer to specific people working within those organisations. Additionally, given the nuanced and critical nature of modern histories of archaeology, these chapters vary in their focus, discussing state formation, politics, law or economics, applying gender theory, or the philosophy of science, or Fleck's theory of thought collectives to illustrate their arguments. The varying viewpoints allow for a more holistic exploration of the instrumentality of informal clusters of actors in the production and mediation of data.

Taking a more explicitly theoretical stance, Milosavljević also considers Fleck's thought collectives, this time in association with a Gephi study, to discuss the development of the culture historical school of archaeology in Serbia during the twentieth century. By examining Fleck's theory in detail, Milosavljević appraises the advantages and disadvantages of using this philosophy in the history of archaeology. As a consequence of its history as part of the socially conservative Yugoslavia and its isolation from Western Europe during the latter half of the twentieth century Serbian archaeology, Milosavljević argues, has a history dissimilar to that of the discipline in the rest of Europe. While these factors led to dogmatism within local archaeological communities Milosavljević looks at how Serbian archaeologists overcame epistemological limitations through informal communication and how this has shaped modern Serbian archaeological thought and practice.

The following two chapters look at the connections and communications between collectors and institutions. Once again informal and fluid networks are the focus of Snead's chapter as he discusses antiquarian communities in the United States during the nineteenth century, looking in particular at the cooperation and competition between antiquarian societies, individuals and the nascent national institutions. Drawing on unpublished documents Snead demonstrates the contrast between local 'amateurs' and their empirical, material-based approach, on one hand, and the more abstract perspective in favour amongst intellectuals, on the other. The Secretary of the Smithsonian Institution, Joseph Henry, attempted to capitalise on the interest in indigenous artefacts by announcing a major report on American archaeology. Lacking sufficient staff to attempt an internally generated report the Smithsonian archaeologists sent out a circular to interested groups and societies. The antiquarian community responded wholeheartedly and hundreds of documents were sent to the Smithsonian, and it is this archive Snead uses to discuss the cultural and social context of nineteenth-century North American archaeology.

De Tomasi also touches on North American collections, but from a very different perspective. In 1889 the Professor of Ancient Topography at the University La Sapienza, Rodolfo Lanciani, was accused of having played an active role in the sale of archaeological objects to North American museums and forced out of his professional positions. While the museums and art galleries of North America and Europe used salaried agents in Rome to acquire materials, many leading scholars, archaeologists and state officials were often called upon to give an opinion on the authenticity and value of these purchases. Lanciani made no secret of his connections with the directors of several North American institutions or his pride at being invited to give a series of lectures at North American universities. Using Lanciani's archived correspondence with General Charles G. Loring, director of the Boston Museum of Fine Arts, de Tomasi discusses the motivation of those who became intermediaries in the Roman antiquities market.

Rome is also the setting for Weststeijn and de Gelder's chapter: they discuss the Dutch excavations that took place in Italy between 1952 and 1958 at the *Mithraeum* under Santa Prisca Church on the Aventine Hill. A combination of favourable political, economic and academic circumstances converged to allow the inexperienced Carl Claudius van Essen, Vice-Director of the Royal Netherlands Institute in Rome, and Maarten Vermaseren, a student working at the Netherlands Institute, the opportunity to direct these excavations. Using a variety of archive sources, Weststeijn and de Gelder emphasise that these successful excavations were as much the result of Italy's reintegration into Europe and the Dutch desire for international cultural status as they were attributable to the work of Van Essen and Vermaseren. Behind the scenes a complex web of negotiations and networks ensured that the Dutch excavation team had the political weight, archaeological expertise, funding and media attention required to successfully undertake the work.

The chapters dealing with individuals are equally wide-ranging while following the central theme of informal communication between antiquarians and archaeologists. De Armond's chapter discusses twentieth-century developments in Czechoslovakian Classical archaeology, the link with European politics and the role played by Antonín Salač. There are few *in situ* Classical remains within the Czech Republic and for most of the twentieth century Prague was far outside the geopolitical centre of Europe, yet Salač managed to create an international reputation as an epigrapher and archaeologist working in Greece and Turkey. It was his connections with French scholars, De Armond argues, that enabled him to be the first Czechoslovakian archaeologist to excavate in these areas. She demonstrates that the encouragement of Salač's work was at least in part a result of French political manoeuvring to promote Czechoslovakia as a bulwark against possible German expansionism. The 1948 communist coup d'état in Czechoslovakia saw an end to Czech–French political relationships and an end to Salač's Francophile leanings.

Gustavsson's chapter similarly examines international relationships between scholars, in this case between the Swedish savant Oscar Montelius and his Italian counterparts. Montelius is best known for his work on seriation and although he is now seen primarily as a 'Nordic' scholar he travelled extensively in Europe and wrote the first work on prehistoric Italy. Gustavsson's chapter reveals how much more there is to Montelius' legacy than typologies of material culture and places his work within a wider, international, scholarly framework of late-nineteenth-century debates about the Italian Iron Age. She places Montelius back within his contemporary and cultural landscape, tracing the connections he made while in Italy and how these networks continued to influence his later work. Using Fleck's theories of thought collectives, allied to actor-network theory, Gustavsson discusses the possibilities and limitations of methodological and theoretical perspectives related to network analysis.

Hansson's chapter continues the theme of northern European scholars involved in Mediterranean archaeology. His examination of the German classical scholar Adolf Furtwängler again focuses on the interaction between scholars, but whereas other chapters demonstrate the constructive results of these collaborations, Hansson discusses how Furtwängler deliberately set himself apart from his colleagues, choosing instead to cultivate contacts within the international art market. During his lifetime Furtwängler was never marginalised as a scholar, but his publication of aggressive criticisms and personal attacks on colleagues resulted in a problematic relationship which then affected the career decisions he made. While the immense quantity of work Furtwängler produced over his lifetime cannot be ignored, Hansson argues that his impact on artefact studies has been overlooked by conventional histories of archaeology as a direct result of his fractious character. Drawing on unpublished archival material Hansson reconstructs Furtwängler's professional networks and work methods.

Arwill-Nordbladh's subject, Hanna Rydh, also encountered problems with her university colleagues, although in this instance it appears that rather than her personality it was her gender, location and theoretical stance that provoked departmental critique. Although based in Sweden, Rydh spent time in France studying Palaeolithic archaeology at the Musée des Antiquités Nationales in St-Germain-en-Laye, near Paris. She published the popular *Millennia of the Cave people* [*Grottmänniskornas årtusenden*] (1926a) to great critical acclaim, but her more scholarly articles were dismissed by her colleagues. These articles, discussing social order, social structure and social organization (Rydh, 1929a, 1931) show how strongly Rydh was influenced by Emile Durkheim's philosophies. Arwill-Nordbladh suggests that Rydh disrupted ideas of gender norms by her presence in Swedish archaeology and then further disrupted academic complacency by adopting alien theories, and as a result was virtually ostracised by her Swedish colleagues.

Demonstrating that geography is crucial not only to the treatment of scholars within institutions, but also to how scholars build their networks to begin with, Sheppard's investigation of James Henry Breasted's early scientific network shows that *where* networks are built is just as important as who is in them. By contrasting the very different relationships Breasted instituted and maintained with Flinders Petrie and Gaston Maspero, Sheppard demonstrates that scientific associations that begin in a space far from a formal institution, such as the field, will maintain that familiarity; whereas connections made in a formal university office will always bear the mark of that decorum. Additionally, these relationships affect the networks produced between scholars and the manner in which information is communicated and utilised. Like others in this volume, this chapter relies more on unpublished correspondence and biographical accounts than on published volumes of scholarship.

Many of the people discussed in this volume worked away from their native countries; in contrast Felix Philipp Kanitz was born in Budapest and became one of the founders and most influential investigators of Serbian archaeology. Mihajlović's discussion of Kanitz and his impact on Serbian archaeology focuses on the latter's role as the central node of a complicated archaeological network. Despite having little, if any, formal training, Kanitz has been called the 'Columbus of the Balkans' and his archaeological work continues to exert considerable authority over modern studies of Roman Serbia. Mihajlović argues that, having been subjected to the frontier colonialism of the Austro-Hungarian Empire, Kanitz deliberately set out to create a network of people from various political, academic, ethnic and socio-economic backgrounds to reflect this environment. It was through these connections that his version of a particularly Serbian archaeology – as opposed to the colonial Yugoslav archaeology – was spread.

Trigg demonstrates the difficulty of finding networks in archives and published works, while discussing the life of Dr Robert Toope. He argues that it is because of Toope's network that we know about him at all. Although he was intensely productive at certain times in his life, and this work was clearly influential on his contemporaries, Toope did not publish his own work, instead relying on the communication and conversational networks that were so common in the seventeenth century. He was a figure who loomed large in antiquarian circles in south-west England, and he appears in the works of those who did publish their ideas, but he failed to make his own ideas public, thereby relegating himself to the dust of the archive. In spite of this, his ideas were and continue to be influential in antiquarian studies.

The diversity of these chapters reflects the current worldwide interest in histories of archaeology; subjects and presenters cover a wide spectrum of periods and places; but all adhere to the contention that the investigation of place, networks and communication in science is indispensable to the study of archaeological history. We said above that these papers had the power to transform the way in which we understand and write histories of archaeology; that may seem like an ambitious claim but we think it is a truthful one. The early histories written by men such as Daniel presented archaeologists in isolation, unaffected by their cultural and social milieu; later histories have addressed this *lacuna* but this is the first volume to argue that place and space affect interpretation, that personality has to be taken into account when discussing the formation of networks and the dissemination of information, and that, while archaeology has always been a communal effort, there is a pattern to that community, a pattern that can be mapped and nodes that can be identified.

The chapters in this volume demonstrate how much more can be said about the history of archaeology, why certain practitioners such as Furtwängler, Rydh and Toope are overlooked by conventional histories, how in order to fit a particular narrative arc archaeologists such as Montelius – and the amateurs involved with the Smithsonian census – have been defined by only a fraction of their work, as has the role played within archaeology by collectors and collecting, an aspect of our history which archaeologists either ignore or view as shameful yet, as de Tomasi demonstrates, is an important strand within our history and one that clearly demonstrates the importance of sites of knowledge and the networks they generate. We cannot possibly understand the significance of archaeologists such as Breasted, Salač and Kanitz unless we are aware of their involvement in international and personal politics: without his French network Salač would not have been able to establish Czechoslovakia's involvement with Classical archaeology. Similarly, in a different international political situation Van Essen and Vermaseren would not have been allowed to excavate the Santa Prisca *Mithraeum*, nor would Kanitz have been able to exploit his experience of imperialism and deliberately create a diverse network that disrupted these colonial boundaries and allowed him to circulate his version of archaeology. All of these examples demonstrate the importance of networks within international political situations, but the personal is also political and, as Sheppard clearly demonstrates, the Breasted who wrote to Petrie was a very different man to the one who interacted with Maspero; physical and social location affect the networks created, just as much as personality and expectations.

This collection is by no means exhaustive in such a broad and deep field as the history of archaeology, and one particular absence is immediately noticeable: with the exception of Hanna Rydh, all the individuals discussed are men. This is especially conspicuous given how many of the chapters are *by* women. This is not a deliberate exclusion, however; as has been extensively discussed within histories of archaeology (Díaz-Andreu and Sørensen, 1998; Smith, 1997, 2000; Roberts, 2005; Sheppard, 2013) and demonstrated here by Hanna Rydh's work, women have been subjected to different social constraints and expectations than men and this is equally true for female archaeologists. Women have often struggled to be involved in archaeology and when they have succeeded their contribution has not always been given the significance it deserves. As Sheppard states in this volume (chapter 9): '[m]any times women were actively involved in scientific networks, running the administrative side of institutional life while the men were in the field. These women are necessary and important parts of these networks, but they are part of the group that tends to be left out of the story.' Nor do any of the chapters discuss the problems faced by other marginalised groups, those whose race or class impeded their involvement with archaeology. Again, this was not a deliberate choice and again, their importance is undeniable, although little studied (Shepherd, 2002, 2003; Lucas, 2001; Roberts, 2005; Quirke, 2010), but their traces are difficult to discern within networks. We know, largely from biographies and anecdotes, that excavation directors employed, re-employed, blacklisted and recommended particular foremen and labourers. Unfortunately, we do not yet know how these workers experienced archaeology, how they felt about their role and how they interacted with their employers.

There are still many questions to be asked and answered, many archives to be explored and it is our hope that this volume provides a foundation that will stimulate other scholars to investigate this valuable field. Without claiming too grandiose a position and purpose for this book, it is hoped that, like Trigger's formative history, it will stimulate debate, investigation and alternative theories.

1

How archaeological communities think: re-thinking Ludwik Fleck's concept of the thought-collective according to the case of Serbian archaeology

Monika Milosavljević

> Both thinking and facts are changeable, if only because changes in thinking manifest themselves in changed facts. Conversely, fundamentally new facts can be discovered only through new thinking. (Fleck, 1981: 50–51; translated by Fred Bradley and Thaddeus J. Trenn).

Introduction

Serbian archaeology offers fertile ground in which to apply Fleck's concepts of thought-collectives and thought-style.[1] To this end, this chapter seeks to delve into Fleck's theories on knowledge production to study how they function in practice in the history of archaeology, as based on empirical data consisting of various texts and citation relations that are used to track a particular thought-collective in a clearer, more visual manner. In doing this, a further aim of this chapter is to introduce new theoretical tools for the history of ideas as well as how they may be implemented as inherent to specific methodological strategies.

Kuhn's concept of a paradigm is limited in its applications since its broad expanse proves too unwieldy to apply to all aspects of a localised phenomenon. Paradigm shift is an appropriate term to describe significant changes that encompass totalities, but not for analysing the specifics of one non-generalised change (Kuhn, 1970). As a matter of consequence, in order to take an initial step into researching the history of localised ideas (such as the history of archaeology), it is necessary to find an approach adequate to understanding the sociology of knowledge production and archaeological epistemology. In this sense, Fleck's concepts are better fitted to taking into account nuances within change that do not correspond to overarching paradigms within larger narrative scopes. Fleck accounts for change as a continual process rather than a single event, and incorporates the social group's role in such changes (Brorson and Anderson, 2001). That said, Fleck's theories by themselves are not theoretically sufficient to cover all issues arising when examining shifts in thought. This chapter's objectives include a retooling of Fleck with corresponding and supporting theoretical sources so as to be able to solidify his theories into an applicable methodological strategy.

This chapter draws on thought-style and (social) network analysis from the actor-network theory (ANT) of Bruno Latour to supplement Fleck in this regard. Because Latour relies on social and natural worlds existing within constantly shifting networks of relationship, the theory can complement the representation of communication between scholars reflecting real-world changes in flux. While ANT does attempt to 'open the black box' of science and technology, it contains no concrete or coherent methodological strategy per se; rather, ANT may be viewed as more akin to a general perspective applicable to understanding social dynamics. ANT's abstract approach distinguishes it from many other sociological network theories. In utilising ANT, it is necessary to do so in conjunction with citation network analysis in order to provide a concrete framework for the methodology itself (Latour, 2005).

Focusing on this constructed methodology in this chapter, we will be able to better comprehend the cultural history school in Serbian archaeology. In doing so, I argue that the school overcame its dogmatic character in the local archaeological community to develop a more democratic academic function. I selected the history of Serbian archaeology as it is a field unique to itself, owing to the extreme difficulty of integrating events affecting Serbian archaeologists with the narrative of the development of European archaeology as a whole. The history of Serbian archaeology has been subject to numerous influences and various shifts in thought over the last century, which distinguishes it from contemporary archaeology elsewhere for its conservatism. A dissection of the development and evolution of Serbian archaeology, therefore, is fruitful for examining how specific shifts in thought occur in non-overarching exceptions to the norm (Palavestra and Babić, 2016).

As a case study, this work directly treats what Kuhn would call a 'paradigm shift': the late introduction of cultural-historical archaeology to Serbia. The objective is not to describe practices in archaeology in detail, but rather to discuss differing theoretical perspectives and tools by reference to Serbian archaeology as it developed over time amid shifts in thought and academic traditions among scholars.

With this objective in mind, this chapter will delve into how it is possible to understand the production of knowledge as a phenomenon of communication within a group when examining this through the prism of a thought-collective. However, as the case study will bear out, while analysing it as such, it is still necessary to take account of the problems of trans-generational transfers of knowledge, and to understand how scientists relate to one another within a network and how to become an authority in a specific scientific field.

Ludwik Fleck in brief

Ludwik Fleck (1896–1961) was a Polish microbiologist, whose studies of the history of medicine and science were written mainly in German and Polish, but remained unnoticed by a wider scientific audience until their rediscovery in the late 1970s (Jarnicki, 2016). Most contemporary scholars now admit that Fleck's contributions are original, even pioneering, in the field of epistemology (Löwy, 2008: 375).

Fleck graduated from medical school at the University of Lviv. From 1920 to 1923, he assisted Rudolf Weigl, famous for his research on typhus. Fleck then went on to specialise in bacteriology in Vienna. From 1925 to 1927, he served as the head of bacteriological and chemical laboratories for the State Hospital in Lviv. He spent 1927 in Vienna, during the heyday of the Vienna Circle.[2] From 1928 onwards, he continued his laboratory practice in Lviv, writing papers on serology, haematology, experimental medicine, immunology, bacteriology, the methodology of science, scientific observations and the history of discoveries. In 1935, owing to his Jewish identity, he was dismissed from the laboratory at which he had worked since 1928. When the Germans occupied Lviv at the start of the Second World War, he was the director of the bacteriological laboratory within the city's Jewish hospital. It was at this time that he succeeded in developing a reliable diagnostic test for typhus, which provided swift detection and isolation in the midst of a typhus epidemic. Fleck was arrested in 1942, along with his family and staff, and they were all deported to the concentration camp at Auschwitz. There, Fleck and his staff were forced to produce a vaccine against typhus for the German forces. In 1944, he was transferred to Buchenwald, where he continued to prepare typhus vaccine. It was only after the Second World War that he received affirmation for his work in the field of microbiology. He became an authority figure in the medical field, which drew attention away from his work on epistemology. In 1957 he migrated to Israel, where he died in 1961 (Trenn and Merton, 1981: 149–53).

Fleck's most significant epistemological papers were published in the 1930s, but became known only with the emergence of constructivist programmes of philosophy and the sociology of knowledge. He has come to be regarded as a standard-bearer in his field, side by side with

Karl Popper or Robert Merton. Further, owing to the influence of Fleck's ideas on Kuhn's *Structure of Scientific Revolutions*, the latter's efforts greatly contributed to the affirmation of Fleck's work after the Second World War (Eichmann, 2008: 26–8; Condé and Salomon, 2016).

According to Cohen and Schnelle, Fleck's scientific work in the field of cognition developed through three phases. First, he slowly migrated from the history of medicine to the history of science with two short essays in 1926 and 1929, in which he began to question scientific reality itself more radically. The main phase of his work on the philosophy of science relates to the publication of his monograph entitled *Entstehung und Entwicklung einer wissenschaftlichen Tatsache* (*The Genesis and Development of Scientific Fact*, 1935). In this work he defined his own theory of cognition. After the Second World War, Fleck's experience called into question the collective basis for scientific work, since he had relied solely on his own experience to develop a typhus vaccine while imprisoned. In July 1960, near the end of his life, his ideas were summarised briefly in an article published in the journal *Science* (Cohen and Schnelle, 1986: x–xi).

When speaking about Ludwik Fleck, his unusual scientific path stands out foremost. The bizarre fortune of his expertise saving him from probable death, his near-invisibility in the philosophical profession, his deferred recognition within that community and the posthumous reception of his work are markers of his unique life, from which his ideas may in part derive their originality. Fleck's contributions to epistemology and science alone bring his genius and tragedy to the fore. Such myths are precisely a type of idea he tended to question and distrust the most: he repeatedly pointed out that the scientific collective must remain the focus when approaching the production of scientific knowledge. Fleck saw the role of an individual as interlocked within a community. He considered claims easily condensed into the form of 'someone discovered something' as vague, since they fail to show any additional dimensions such as social context, social networks and the understanding of the claims such statements make. To establish that 'someone discovered/recognised/pointed out/dug up something' is possible only when the basis of the existing knowledge is already known. This is to say that a conclusion may be reached only within a particular cultural ambience, thought-style or thought-collective (Fleck, 1986 [1947]: 134–40; Weissmann, 2002: 112–13; Condé and Salomon, 2016).

To summarise, the concept 'thought-collective' represents the idea of a community of people in constant intellectual interaction exchanging their ideas. The members of the thought-collective accept specific ways of perception and thinking and tend to share a style of thought that gives birth to 'the real explanation'. Even though a thought-collective is a group of individuals, crucially it does not form by simple addition of people or by their actions within a single frame (Fleck, 1981: 41). It instead forms by a group dynamic, which imposes a collective manner of thought from which individual thought shows no variance.

Re-thinking knowledge production on a Fleckian basis

In Fleck's manner of thinking, the transformation of an idea originating from interpersonal communication is key. The backbone of a thought-collective lies within communication which values three main processes: understanding as well as disagreeing; diverse understanding of the same phenomena; and linguistic articulation of ideas. While different thought-collectives can research or describe the same subject, communication between thought-collectives can be very difficult. Since Fleck regarded thought-collectives as working not only in science but in the arts, religion or politics (to name a few areas), he put forward astronomers and astrologists as an example of such impossibility of communication between two thought-collectives. Even though both collectives reach conclusions by observing celestial bodies, their styles of thinking are incommensurable. By defining this as the problem, Fleck does not underestimate the significance or the position of science, but allows irrational elements in scientific thinking to be susceptible of analysis. Furthermore, in his view there are differences between scientific and non-scientific thought-styles, which relate to the density of interactions between participants in thought-collectives. Scientific communities are characterised by a high density of social interactions; as a consequence, scientists tend to produce consensual and homogenised knowledge (Löwy, 2008: 382).

It is difficult to overlook the social structure of scientific communities, even when considering only the formal aspects of their actions. Simply examining the distributions of work, cooperation, co-authorships, aspects of technical support, the exchange of ideas and controversies within the scientific community will call attention to this. Moreover, groups and hierarchical positions within the same community can be distinguished through observing participation in meetings, congresses and professional journals as well as different approaches to professional training, field experience and academic exchange (Fleck, 1981: 38–44).

However, the question here is how thought-styles are formed and how thought-collectives function. Ideas pass from one person to another and produce slightly different associations. In Fleck's opinion, one can never speak of an absolute understanding of an idea. After a few exchanges about the interpretation of a given phenomenon, almost nothing remains of the original idea. Given this flux, what then is the understanding that is kept in circulation? By exchanging ideas within a community, key subjects are improved, changed, reinforced, simplified, ultimately influencing the formulation of concepts, customs and habits within a community. When, after several rounds of exchange, the altered ideas return to their 'originator', who is, per se, changed by the pulsations of the process of exchanging thoughts, that 'originator' might perceive the newly made ideas as their own, that is, containing nothing more than the initial idea did. The key specific of a thought-collective is that we see with our own eyes, but perceive through the lenses of the community we belong to. We know whether we are able to see what it is to be collectively acceptable (Fleck, 1981: 42).

Some thought-collectives last only a short time, but those with an organization and structure able to last for several generations often resemble religious movements, and consolidate authority and influence in a similar fashion to 'national traditions'. Long-lasting thought-collectives are intertwined with the institutions through which they induct younger generations, by virtue of an educational system and rituals following the induction of new members into a community. When a thought-collective grows large, it definitely becomes a more widely-extended and sophisticated system. It consists of a small circle of experts (an esoteric circle) from which, in part, knowledge originates; and a group of scientists in a wider circle (an exoteric circle) who are under the influence of the group's style of thinking, but do not play an active role in formulating and changing this. The central figures, or members of esoteric circles within scientific communities, are equivalent to preachers to whom others extend trust. It is interesting that Fleck argues that popular and textbook science, always slightly simplified and seemingly convincing and well based, reinforces belief in the objectivity within the scientific community. Hence, it functions as a loop: the scientists preach to the broadest public possible, who in turn consider their statements as relevant and express respect for them, and in turn the scientists see public desire as overlapping with their own work. Within the inner structure of a thought-collective, Fleck distinguishes the following subgroups: 1) the group preceding the thought-style, working practically on a given problem (the vanguard); 2) the official community; and 3) the stragglers (Weissmann, 2002: 110–11; Škorić, 2010: 350).

Highlighting the characteristics of a thought-collective allows further discussion of the basis for it. First, solidarity develops within members of a thought-collective, a mother scientific group, comprising colleagues. The group develops disdain for the members of other thought-collectives. They are strangers, believing in other gods, using unfamiliar words and unreliable concepts. According to Fleck, emotions play a large role in the function of scientific communities. In a researcher, they often inspire dedication through participation in a given mission and accentuating the significance of initiation into the research circle. It is possible to distinguish democratic thought-collectives (the most common interpretation of the character of the scientific community), in which every member is encouraged to study and advance, from dogmatic collectives, which develop dogmatic ways of thinking, basing rules of conduct on some mythical figure/founder/saviour from the distant past. Everyday life in the latter type of community has a reinforced, ceremonial character and access to esoteric circles is well guarded. Within these circles there is no room for fundamentally new ideas – only a more precise following of existing principles. A thought-collective is more likely to succeed when research is conducted under explicit social pressure; that is, if researchers work long enough on a certain problem and receive sufficient material support (Fleck, 1981: 98–115).

Fleck opened *Genesis and Development of a Scientific Fact* in 1935 with the questions what is a scientific fact? How is it created and developed? In his view, the sanctification of facts by itself produces an extreme passivity in the scientific community, as the reality of facts is regarded as completely independent of the scientists establishing them. Questioning this should not induce scepticism, but rather revive the dependency of cognition on the thought-collective. Through understanding this relationship, it is possible to understand when and how facts change (Fleck, 1981: xxvii–viii). Once a thought-collective is established, the scientific observations that stem from it become strictly defined by the collective's limitation to the boundaries set within its established viewpoints. The thought-collective therefore actively resists all contradictions of its established world view, through several distinct phases:

1 Contradicting the system is incomprehensible.
2 Tending to ignore anything that does not fit within the system.
3 If any aberrations are then noticed, they either remain a secret, or obvious efforts are made to explain them in such a way as to bring them within the system in a particular way.
4 Despite any justification for contradicting standpoints, the individual starts noticing, describing and illustrating those circumstances that fit closest to current understandings, to participate in the meaning within the terms accepted by the thought-collective (Fleck, 1981: 27).

Perceiving a new fact is not possible unless a scientific community changes its thought-style, or at the very least a change is indicated. In the process of changing, small transformations, misunderstandings and mutations of ideas occur in which constant interactions play important roles. It is impossible to learn and adopt something radical in a simple and swift way. In addition, the triggers for change can come from completely unexpected directions, such as from proto-ideas (Škorić, 2010: 344–5).

The concept of proto-ideas enables us to understand trans-generational processes, the development of ideas on the vertical scale of the heritage of a discipline, such as archaeology in Serbia. Fleck regards proto-ideas as rudiments of contemporary theories, indicating that facts are always established step-by-step, starting as unclear proto-ideas which are neither correct nor incorrect. Considering that the task of epistemology is precisely to discover this transformation of ideas over time, he emphasised the significance of understanding proto-ideas, pointing out specifically that the detection of irrational elements in obsolete explanations could help scientists to better contextualise their own scientific knowledge (Rotenstreich, 1986: 161–76).

According to Fleck, proto-ideas constitute a significant part of our socio-cultural heritage and, at certain moments, present the thought-collective in the process of cognition. Conversely, he accepts no thesis about scientific knowledge being cumulative; rather, science is a continuous change of thought-styles that develop over time, are sociologically conditioned and interact mutually. The dynamics of this structure generates the development of science but development can be taken as neither progressive nor evolving. New knowledge ensues and old knowledge is lost, not through progress but through certain problems losing relevance to a thought-style. Unlike Kuhn, Fleck does not speak of revolutions (Brorson and Andersen, 2001: 123). Fleck notes that scientists are not aware of changes. Certain ideas have a longer lifespan because they present inspiration to newer thought-styles then are reinterpreted in accordance with changes in a thought-style (Škorić, 2010: 346).

Archaeological communities (think) as thought-collectives

There is a general consensus that delays occur in the adoption in peripheral environments (such as Serbia) of archaeological concepts originating in Western European. This would imply (falsely) that in general the development of archaeology follows the same uniform, unilineal sequence of paradigms: culture-historical, processual and post-processual (Babić, 2014; 2015; Palavestra and Babić, 2016: 317). However, the concept of paradigm and paradigm shift is not applicable to Serbian archaeology as far as Kuhn is concerned, since it is too unwieldy to apply in all aspects on the local level (Kuhn, 1970). As a consequence, to constitute a first step into research of the history of ideas in archaeology, it is necessary to find an approach adequate to understanding the sociology of knowledge production and archaeological epistemology. In this sense, I will focus on the cultural history approach in Serbian archaeology, and discuss how it overcame its dogmatic character in the archaeological community. In order to comprehend this shift within Serbian archaeology, it will be necessary to adopt Fleck's concept of the thought-collective as a novel tool in the understanding of networks of communication and the production of knowledge within archaeology. However, his concepts have demonstrated their limited scope for understanding trans-generational knowledge transfer, which necessitates further examination and reflection upon these theories, to adapt them to be more applicable.

To improve Fleck's definition of the thought-collective, it must be developed further, to be utilised as a tool through the creation of a research programme for this specific case study – which will be based on four distinct steps. The first and foremost is Fleck's understanding of how a thought-collective works (Fleck, 1981), presenting a unit for studying a horizontal cross-section of the history of archaeology. Secondly, to connect different generations of archaeologists, it is necessary to strengthen Fleck's thought-collective through the prism of Karl Mannheim's concept of 'generation'. Mannheim asserted that a generation is determined by the similarity of a social location, primarily through his understanding of how generational experience is 'stratified'. He examined how knowledge is transferred between 'generations', how the hierarchy of research questions is forgotten by later generations, how different groups (thought-collectives) establish themselves within a single generation and the precursors of the generational style. Mannheim's ideas serve to connect the horizontal cross-sections of the history of ideas in archaeology (Mannheim, 1952). Metaphorically speaking, when Fleck's and Mannheim's ideas are combined, a horizontal and a vertical axis are achieved. However, even this graphic representation does not embrace the full complexity of the transfer of ideas occurring in Serbian archaeology.

The third step in the creation of an applicable research programme is to include the actor-network theory of Bruno Latour, who views the production of scientific knowledge as occurring via relations within a network. From this point of view, the intrigues, dialogues, agreements and disagreements, as well as both formal and informal discussions within the scientific community, can be visualised as a comprehensive unit to be analysed. Although ANT utilises a wide vocabulary in order to surmount such a complex issue, its vocabulary is frequently misused and misunderstood in its application. As to avoid this pitfall, this case study has disregarded Latour's expansive *œuvre* in favour of concentrating on his early work, particularly the birth of social constructivism in the post-Kuhnian philosophy of science. Latour's initial approach, demonstrating that scientific 'facts' are not an out-there 'substance but fabrications' emerging from social interactions, is crucial in understanding the microstructure of academic networks (Latour, 2005).

Last, but not least, are the points of intersection; that is, strongly networked knots within a network or authorities within a scientific community. In the field of archaeology, Tera Pruitt addressed this issue on a Foucauldian basis in her doctoral dissertation *Authority and the production of knowledge in archaeology* (Pruitt, 2011).

Introduction of culture-historical approach into Serbian archaeology

Let us therefore look more closely at one particular example of Serbian archaeology. During the first half of the twentieth century the discipline was predominantly marked by the ideological domination of a single authority who actively suppressed scientific debate, but also the development of new scientists stemming from emerging generations and dissenting interpretations of the past. This authority was Miloje M. Vasić, a classical archaeologist educated in Berlin and Munich in the late nineteenth century (Palavestra, 2012; 2013). He defended his PhD under the supervision of Adolf Furtwängler, the so called 'Linnaeus of classical archaeology' (Hansson, 2008: 19–23; and 2014) who has been described as described as 'more feared than loved' (see chapter 7). Yet Furtwängler's influence would bring fruitful results: Vasić was to become the ultimate archaeological authority in Serbia, resulting in an era of his absolute domination over the discipline in Serbia which lasted throughout the first half of the twentieth century.

Starting in 1908, Vasić began to systematically excavate Vinča, a multi-layered, prehistoric archaeological site of great importance on the shores of the Danube near the Serbian capital of Belgrade. The excavations were occasionally interrupted by war over the following decades. From the first reports of the excavation, Vasić began to interpret certain evidence as culturally influenced. In his opinion, these influences had spread north-east from the Aegean region, and Vinča was proof of this. To him, the site dated from the Bronze Age and had been settled by Aegeans along with autochthonous locals. Ultimately, in 1934, Vasić came to alter his interpretation, concluding that Vinča had been an Ionian colony on the Danube dating from the sixth century BCE. He staunchly defended this faulty thesis until his death in 1956, even in the face of overwhelming archaeological discoveries and interpretations that solidly proved the falsehood and unsustainability of his theories (Palavestra and Milosavljević, 2015: 322).

After the Second World War, and then Vasić's death, the establishment of working interrelationships between Serbian and other Yugoslav archaeologies led democratic tendencies to develop within collectives (Novaković, 2011; Milosavljević, 2015). It was only after the war that his interpretation began to be criticised by some of his students, who included Josip Korošec, Milutin Garašanin, Draga Garašanin, Alojz Benac and Vladimir Milojčić. This period has been called by the present Serbian archaeological community a paradigm shift, in which a cultural history approach in Serbian archaeology became established following the recognition of Vinča as a Neolithic site (as it is), not a supposed Ionian colony. The shift by itself was not the driving force behind the change, but rather its catalyst (Palavestra and Babić, 2016: 324).

The work of Gordon Childe was well known to Vasić even prior to the 1920s, so much so that Childe had come to Vinča officially in 1926 to speak with Vasić. Childe reportedly considered Serbia to be one of the most significant areas for improving Europe's understanding of prehistory (Nikolić and Vuković, 2008: 39–86). It must therefore be asked why, in light of this familiarity, a cultural-historical approach already systematically established outside Serbia had to wait another thirty-odd years to be introduced into Serbian practice.[3] Put more bluntly, what must already be established before new thinking can emerge, let alone its application within a knowledge community?

Gordon Childe and C. Daryll Ford, his friend from Cartwright Gardens who later became professor of anthropology, travelled together for six weeks throughout Yugoslavia, Romania and Hungary in 1926, gathering new data. Special attention was focused on personally checking the stratigraphy of sites such as Vinča when Childe visited Vasić's excavations near Belgrade. During that period, *The Dawn of European Civilisation* was printed. By the September of that year, *The Danube in Prehistory* had been finished, in which the Vinča site was recognised as key for the study of European prehistory and the Danube as an extensive natural highway across the European continent, the principal route along which civilisation had been diffused from the Near East. The justification for Childe's chronology was his synchronising of Vinča I with Troy II, based upon his economic perspective, which he would go on to use for the remainder of his life, helping him change the face of European archaeology (Trigger, 1980: 56–60; Green, 1981: 55–6).[4] Vasić was impressed by *The Danube in Prehistory*, mostly because of the Aegean-Danubian parallels cited in it. His only point of contention was the dating of the Vinča site. What is most pertinent is that mechanisms of cultural change were to be found both in Childe's and Vasić's work, for all that the latter's ideas about Aegean influences predate those of the former (Palavestra and Babić, 2016).

Compared to the state of Serbian archaeology under Vasić's domination before the Second World War, the second half of the twentieth century began with a greater number of once-marginal figures and young people being included within the archaeological community. Such a change was possible owing to the role that Miodrad Grbić played, educating young colleagues and establishing international contacts, which allowed continual access to new information in the field. He held the post of part-time director in the Serbian Ministry of Education under German-occupied administration during the Second World War. As a consequence, he initiated a controversial two-year course at the National Museum of Serbia, by means of which he educated young college students on archaeology, the history of art and museology, showing a great number of them approaches that differed from Vasić's. As Lidija Ham-Milovanović has pointed out, '[i]t was a unique opportunity for new generations growing up at the time of the occupation because Belgrade University was closed and did not enrol new students' (Ham-Milovanović, 2009: 121–2). A wide spectrum of topics had occupied the attention of archaeologists in Serbia before the Second World War, and the interpretations found in the works of Miodrag Grbić were among them, alongside Vasić's standpoint. Grbić's interpretations are of extreme importance owing to the eventual role that the course, organised under his guidance, would play in the history of Serbian archaeology (Bandović, 2014: 629–48). However, like many others, Grbić was socially marginalised after the war because he had refused to distance himself from any form of cooperation with the German-led administration. The thought-collective headed by Milutin Grašanin introduced the cultural history approach to Serbia after the Second World War.

While the pre-war generation of Vasić's students established a community after the war which could be called a thought-collective, the core consisted of pupils who attended Grbić's course and emerged as a collective of resistance to the ideas of Vasić. The majority were students who had begun their studies of the classics with archaeology at the Faculty of Philosophy of the University of Belgrade in the 1930s. Aleksandar Palavestra has described this group of Vasić's students and Grbić's co-workers, comprising Josip Korošec, Alojz Benac, Milutin Garašanin and Draga Aranđelović-Garašanin, as a Fleckian thought-collective (Palavestra, 2013: 685). The oldest among them, Josip Korošec, left for his doctoral studies in Prague, where he earned his degree under Lubor Niederle at Charles University in 1939. Undergraduates who studied in Belgrade before the outbreak of war were later to complete their doctorates at the newly founded Department of Archaeology at the University of Ljubljana (Slovenia), headed by its founder, the same Josip Korošec (Milosavljević, 2015: 172–80).

Milutin Garašanin was one of these students. He sought to complete his doctoral studies under Korošec as he had a severe disagreement with Vasić over the dating of Vinča. This was a clear sign for Garašanin that doctoral work on a Neolithic topic could not be defended in Belgrade, owing to current academic biases (Babić and Tomović, 1996: 20).

Following the example set by Grbić and armed with the doctorate obtained from Niederle, the independent position of Korošec therefore made it possible for an entire generation to escape from Vasić's shadow. Miodrag Grbić had also defended his doctoral dissertation 'Pre-Roman bronze dishes in the region of Czechoslovakia' under the supervision of Lubor Niederle, but he had done so in 1925 (Gačić, 2005). In any case, no further anecdotes are needed about the fraying of relationships within the archaeological community during this period, but it is in such a context that the flow of knowledge transfer is determined alongside the disciplinary continuum. If such problems are examined with a view to identifying Mannheim's generations in the sociology of science, it is useful to note that the term 'generation' does not refer to a specific social group. Mannheim compares the term 'generation' to 'class'; he states that the force binding the members of a generation is the same as that binding a class – shared social location. Mannheim also states that the members of a generation share a layering of experience in social life, though not all members of a generation may experience the same events even when they live contemporaneously with one another. Subsequent experiences are usually assigned meaning based upon the first set, either in confirmation or as a negation of those first experiences – from this starting point any two alternating generations can have completely different primary orientations (Mannheim, 1952).

For instance, Milutin Garašanin, a key representative of Serbian archaeology after the Second World War, belonged to a generation which could complete its undergraduate studies under Vasić, the museum course under the guidance of Grbić and the first intertwining excavation of Yugoslav archaeology in Ptuj. He lived through the same formative layering as other students of his generation. Vasić, however, was of another generation, and perceived new developments at the end of his career as contrary to his 'stratification' of experience, as he defined them within his own generation's thought-style (Novaković, 2011: 396–8). As Aleksandar Palavestra and Staša Babić summarise:

> Thus the concept of culture groups, around which the culture-historical paradigm is mainly built, entered Serbian archaeology indirectly and from various directions, and was not understood in the same way by the archaeologists of successive generations, Miloje Vasić, Miodrag Grbić and Milutin Garašanin. Though modestly present in Serbia in the 1930s,

the paradigm came to be dominant only in the years after World War II (Palavestra and Babić, 2016: 324).

Tracking the thought-collective

The history of the discipline of archaeology is generally presented as an uninterrupted chain of key authorities and their ideas; yet great archaeologists and great discoveries have not been uncontroversial. The development of the archaeological community in Serbia, for instance, can be analysed by reference to the critiques and reviews published in professional journals, the content of which may be seen as casual and unimportant reading material; but it is, in fact, important in analysing what science contains. However, such material provides succinct views of what is considered important in a given context. Most commonly, reviews retell a book or a journal article, although occasionally the authors of a review sharply criticise or explicitly stress the significance of a particular publication. The points at which the reviews leave the tracks of unvaried summarising are often important indicators for contextualising ideas that are crucial for the thought-collective. Such points in the text either represent an underpinning to the research path or set boundaries between 'good science' and tangential diversion. Following Latour's actor-network theory, the importance of networks as social connections (Latour, 2005), as well as structures that support and propagate facts and archaeological theories, comes to light. Through examining communication networks among archaeologists (or thought-collectives, in Fleck's terms) – their emergence, support mechanisms and what disrupts them – it is possible to gain a richer understanding of how theories travel. Furthermore, archaeological methods and conventions, clearly visible in reviews which produce data in a particular context, do not stand alone. They need to be supported by a network of recognised authorities, hence the need for publication and scholarly exchanges.

The question that must be then posed is what network was central to the effort of establishing a cultural history approach within Serbian archaeology inside the newly formed Yugoslavia of the time. To wit, what narrative strategies were used to achieve that goal? Also, how did archaeological networks and citation practices function in this particular context?

To answer these questions, a critical analysis of discourse found in reviews from the prominent Serbian archaeological journal *Starinar* (*The Antiquarian*) from the years 1950 to 1960 will be carried out in order to better understand the changes experienced in the archaeological community of that time.

In the first issue of the new series of *Starinar* (1950), Garašanin reviews the fourth edition of *The Dawn of European Civilisation* from 1947, a quarter of a century after the first edition. He considers that the basic concept of the book, guiding the author in the treatment of material culture, is different from the common understandings of prehistory in Europe. Nonetheless, Garašanin deems this approach more realistic and more acceptable, as it is based on the social-economic foundations of prehistoric society. Certainly, he is more interested in how Childe's attitude towards the question of Vinča culture has changed in this book, from that expressed in *The Danube in Prehistory* (Garašanin, 1950: 257).

In *Starinar* III–IV (1955), a review written by Vasić of *The Dawn of European Civilisation* appears, but of the French edition of the book published in 1949. Vasić's criticism is sharp and foremost refers to Childe's understanding of Vinča. Vasić states that the book is a compilation, completely in need of a rework (Vasić, 1955: 233). Opposing Vasić's position, in *Starinar* (1959) Garašanin once again reviewed the sixth, updated edition of *The Dawn of European Civilisation*, published in 1957. He stated that Childe's work is regarded as a classic for prehistorians (Garašanin, 1959a: 392–3). In the same issue, Garašanin writes an obituary of Childe. He notes that Gordon Childe was tireless at his work, especially in persistently following new research and studies. It seems that the famed archaeologist would be seen as an antithesis to Vasić, since 'he had always and gladly accepted discussion, possible objections and remarks, ready to openly admit fallacy and accept corrections concerning their legitimacy' (Garašanin, 1959b: 446).

It bears repeating that attitudes towards Childe's work can be viewed as an indicator of the general direction in which archaeology flowed within the post-war generation of archaeologists in Serbia, as headed by Garašanin. In a certain sense, Childe superseded the negative experience that Vasić represented. In the local application of general trends in archaeology, the Serbian cultural history approach formulated after the Second World War was substantially linked to Central European archaeologists such as Gero von Merhart or Richard Pittioni (Novaković, 2012, 151–71), as well as the 'late' Childe – that is to say, his understanding of the culture implemented in Yugoslavian/Serbian archaeology after the Second World War could be compared to those who pointed up changes in material culture which do not necessarily demonstrate change in ethnicity (Novaković, 2011: 440–50; Raczkowski, 2011: 201). This 'late' version of Childe's thinking began in the 1930s, when he overtly discarded the connection between race/ethnicity and archaeological culture based on ideas borrowed from Soviet archaeology (Patterson and Orser, 2004: 9). However, when the main weapon of the cultural-historical school of thought was questioned in the West after the Second World War, Gordon Childe also simultaneously became a landmark and a yardstick for the systematic scientific approach in Serbian archaeology (Babić, 2014: 286–7).

Visualising the thought-collective

The example of Serbian archaeology has hereto been drawn upon to demonstrate the experimental use of Fleck's concepts of thought-collectives and thought-style. This chapter discusses a Fleckian theoretical background for the history of archaeology in Serbia based on empirical data consisting of various texts, as well as relations between citations that are used to track a thought-collective in a clearer, more visual manner. To this end, the following section will delve more deeply into the development of this methodology, seeking to represent a thought-collective visually by mapping the function of relation networks. The purpose of this methodology has been to apply Fleck's ideas to the history of archaeology proper.

Social network analyses have proven to be useful as a formal concept when applied critically to trace the domestication and adaptation of ideas, methods, and techniques by thought-collectives. Network analysis is not a single, homogeneous method, but rather incorporates every formal technique that visualises or analyses the interaction between nodes. According to Tom Brughmans, a formal network is a set of nodes as well as the ties connecting them (Brughmans, 2013; 2014). A citation network analysis is a useful approach to explore general trends in academic influence; co-citation networks are a fruitful indicator, in particular of clusters of papers that deal with related topics. By carrying out a citation network analysis (Waingart, 2015: 201–13), the connections made by co-citations among key authors in Serbian archaeology during the second half of the twentieth century can be shown, from which highly indicative results are obtained. Of course, citation analyses have abundant methodological issues, particularly when this technique is applied in this primarily quantitative form. Whatever the issue, citation is a process in which the author creates private symbols for certain ideas that they use by citing a text. Private symbols easily become 'standard symbols' for a particular group of researchers (in the frame of an 'invisible college') (Díaz-Andreu, 2008: 126–7). Often, citation not only refers to the author being cited, but, for a certain thought-collective, links that author to a referent representative. Since citation depends not only on the object of a work but also on the individual who is citing and the social context within which they are working, it is therefore necessary to keep context constantly in mind (Škorić, 2010: 266–75).

The question of the 'Illyrian' or Palaeo-Balkan past was one among a number of common topics for archaeologists in the former Yugoslavia, primarily within the thought-collective headed by Milutin Garašanin (Džino, 2014; Babić, 2014; Mihajlović, 2014). This topic has been selected for analysis because the 'Illyrian' or Palaeo-Balkan past is a matter of identity for Yugoslav and post-Yugoslav archaeologies (Gori, 2014: 300). It serves as both an apt example and a rich source of sampling for such an analysis. A co-citation network is helpful when needing to gain both a clearer picture of the discussion carried per se within a field and better insight into how the topics of disciplinary conversation interconnect or fail to do so. Since the use of history and archaeology is susceptible to the self-interpretation triggered by terror of the *Zeitgeist* in academia, examining the interaction of a collective whole of inter-citation helps to access the core of thought relevant to a given period. One could say that thought-collectives are detected inductively using this technique and, as it is in essence descriptive, it works better as a tool to help clarify ideas about the field than to prove or disprove hypotheses. It is important to bear in mind the object of representation within this approach: co-citation networks generated from limited, selective material. By necessity, this underscores the fact that it does not provide a complete picture of the field; rather, co-citation analysis is an apt method for identifying who was most influential in Serbian/Yugoslav archaeology (during a particular period) (Gmür, 2006; Waingart, 2015).

I have selected seven of the most prominent texts (according to how often they were used for teaching) on 'Palaeo-Balkan tribes' published between 1950 and 1990. The scientific texts in this sample were written by four archaeologists: Milutin Garašanin (1964 and 1988), Alojz Benac (1964 and 1987), Dragoslav Srejović (1973 and 1979) and Branko Gavela (1971). The visualisation of the co-citation network was prepared using the Gephi platform, designed for visual representation in research into networks and complex systems (https://gephi.github.io, visited 13/05/15). The resulting network comprises 297 nodes and 414 edges (Figure 1.1). The analysis processed 1,118 ties, which presents a modest span of research though still relevant for visualisation.[5]

Based on this sample, the conclusion is that Garašanin was the most intertwined or central figure in the thought-collective to which he belonged (Figure 1.2 (a)). Moreover, a key connection lies between Garašanin and Benac (Figure 1.2 (b, c, f)). The number of elements of bibliographic coupling for these two authors is highly significant (Figure 1.2 (f)). What is most salient is the weak intertwining of Srejović (Figure 1.2 (d)), who was one of the most important figures in Serbian archaeology during the second half of the twentieth century owing to his great discovery of Lepenski Vir (Novaković, 2011: 397–8). One plausible reason would be that he was central to another thought-collective,

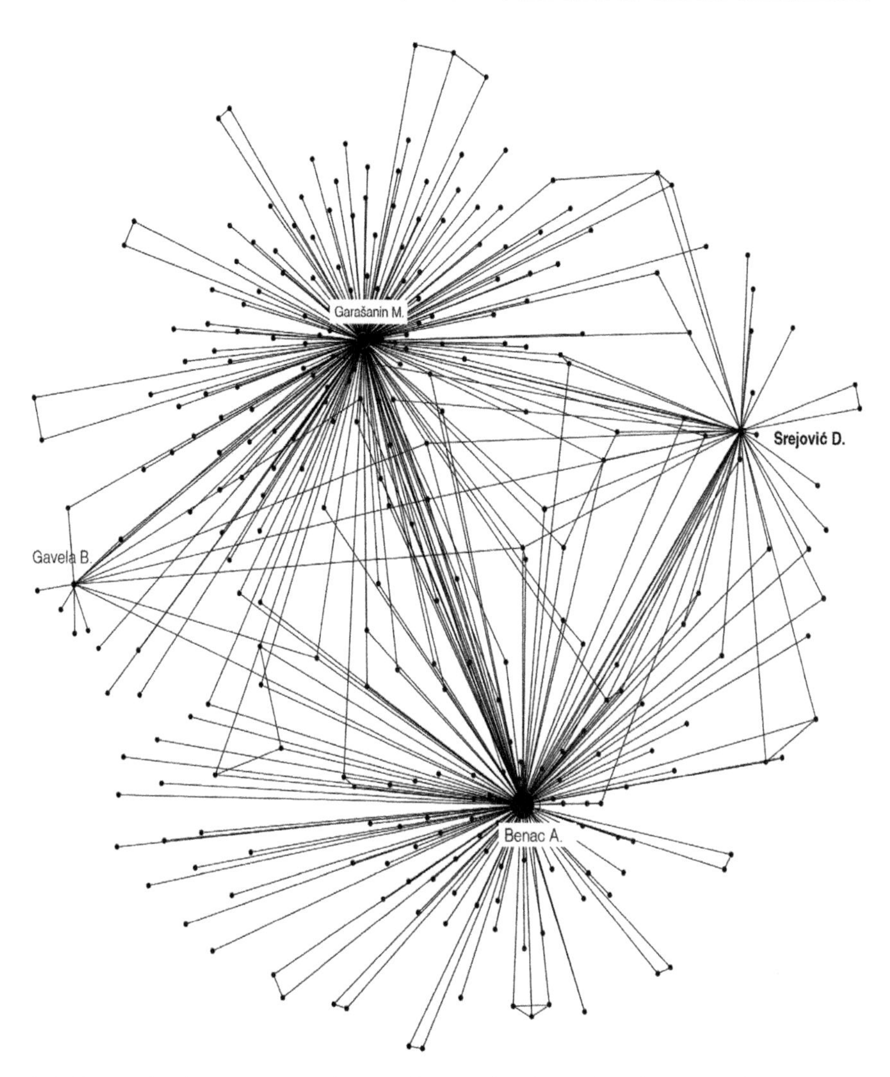

1.1 Network of co-citations of scientific texts by the archaeologists Milutin Garašanin, Alojz Benac, Dragoslav Srejović, and Branko Gavela. Copyright © Monika Milosavljević. All rights reserved and permission to use the figure must be obtained from the copyright holder.

which was opposed to introducing a culture-historical approach into Serbian archaeology. Despite their disagreement, changes in Serbian archaeology during the 1950s occurred as a consequence of communal agency among Yugoslav archaeologists, headed by a thought-collective constituted by Josip Korošec from Ljubljana, Alojz Benac from Sarajevo and Milutin Garašanin from Belgrade as well as their local networks of archaeologists.

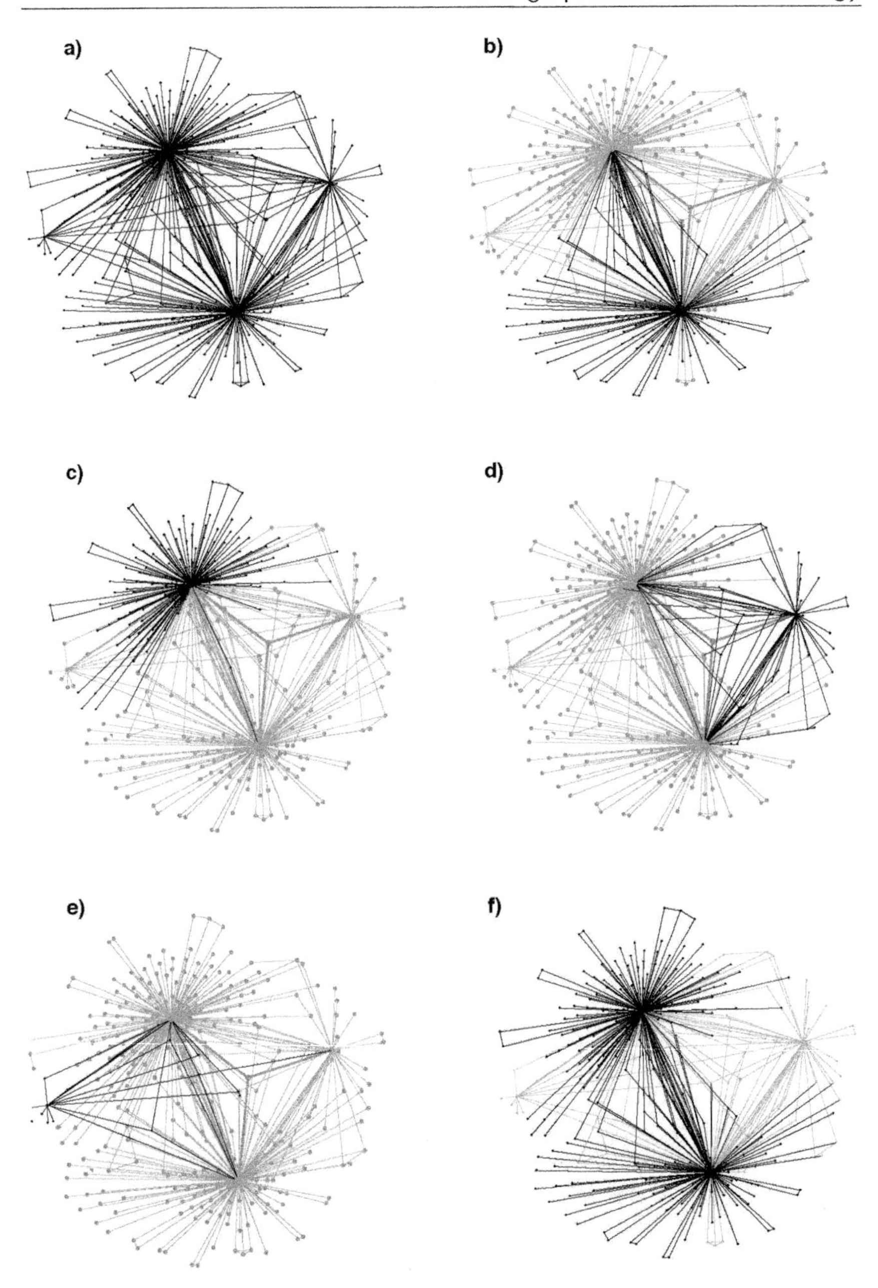

1.2 a) Network of co-citations; b) Alojz Benac's influence in the network; c) Milutin Garašanin most intertwined in the network; d) weak intertwining of Dragoslav Srejović; e) Branko Gavela's influence in the network; f) a key connection between Milutin Garašanin and Alojz Benac. Copyright © Monika Milosavljević. All rights reserved and permission to use the figure must be obtained from the copyright holder.

If the history of archaeology is relevant to science, it is essential for it to develop its own theories and methodologies. This fact becomes clearer when Kuhn's approach is considered: it is far too simplistic to encompass the complexities of academia and academic research (Kuhn, 1970: 10–22). To counteract Kuhn, this chapter has undertaken a critical application of the theories of Ludwik Fleck on knowledge production to explain how a cultural history approach was introduced and thrived in the field of Serbian archaeology. Through a co-citation network analysis Fleck's concept of the thought-collective and the ways it functions has been demonstrated here to be germane, principally because no revolution took place, but rather a change in thought. The process of change examined was protracted and occurred under complex mitigating circumstances; and it is highly significant that the strain which thought-collectives underwent led them in a single direction. As a way to analyse the consolidation of new knowledge within the collective, it has been extremely important to be able to select an adequate sample that reflects already established and accepted forms of knowledge taught within the collective. However, to better gain insight into the actual changes within the collective as they interacted with one another, this chapter has shown that network and co-citation analyses serve well in establishing patterns within such changes.

Notes

1 The research presented here was undertaken for the purposes of project No. 177008, funded by the Ministry of Education, Science and Technological Development of the Republic of Serbia. I am grateful to Vladimir V. Mihajlović, Aleksandar Palavestra and Staša Babić for providing useful comments and criticism. Responsibility for errors is mine alone.
2 The Vienna Circle was a group of philosophers who met regularly in the 1920s and 1930s at the University of Vienna, chaired by Moritz Schlick. The group was highly active in advocating new philosophical and epistemological ideas in the field of logical positivism.
3 It should also be noted that Childe's works were readily available in the library of the Faculty of Philosophy in Belgrade, and that Vasić had recommended them to his students. As a curiosity, on one page of Vasić's copy of *The Danube in Prehistory* there are seventeen exclamation marks!
4 Childe recognised the presence of the *spondylus* shells in the Vinča I stratum and interpreted them as evidence of Neolithic trade, possibly in return for cinnabar ore. As this was not Childe's original interpretation, it very well could have been prompted by something Vasić had noted during the former's visit in the summer 1926 (Trigger, 1980: 59; Palavestra, 2013: 700–701).
5 Automated citation indexing has changed the way that citation analysis research is carried out, allowing data to be analysed for large-scale patterns; unfortunately, this was not possible within the scope of research for this chapter. Consequently, bibliographies have been extracted manually.

2

Circular 316: archaeology, networks, and the Smithsonian Institution, 1876–79

James E. Snead

Introduction

On November 26, 1874, Chicago's *Daily Inter-Ocean* ran a story featuring William Berridge, resident of the town of Pecatonica, Illinois, who had discovered an ancient burial while digging a well on his property. 'After he had got down about ten or fifteen feet,' the article noted, 'his spade struck into something hard, which turned out to be a human skull.' Berridge, it seems, took little interest in his finds, but word spread quickly 'and the people flocked from all parts of the country to see these WONDERFUL REMAINS.' (*Daily Inter-Ocean*, November 26, 1874).

Late-nineteenth-century American newspapers like the *Daily Inter-Ocean* were abuzz with stories of antiquarian discovery. The *Memphis Avalanche* chronicled remains 'found on Mrs. Imogene Beaumont's place, situated on Lake Cormorant, De Soto County, Mississippi' (*New Orleans Times*, July 12, 1874), and dozens of similar accounts were published. Collectively, such reports demonstrate that antiquities were a common element of American rural life, engaged with interest and curiosity.

Such evidence for the popular appeal of archaeology in the United States during the decades following the American Civil War also exposes the chaotic state of 'professional' practice in that era. Diverse communities of interest flourished in the American hinterland. Antiquarian entrepreneurs amassed collections and dealt artifacts through extensive networks; local and regional societies pursued fieldwork, published reports, and promoted cultural achievement. Competition between these communities and networks was commonplace. In particular, structures of inquiry – access, interpretation, authority – were the subject of vigorous debate.

Leaders of nascent national institutions understood the popular appeal of American antiquities, and made use of it to establish control over archaeological practice. The Smithsonian Institution – in particular, its founding Secretary, Joseph Henry – played a central role in this process of professionalization. Although his own research concerned electromagnetism, Henry promoted archaeology, publishing five archaeological reports in his first decade as Secretary via the new *Contributions to Knowledge* series (Whittlesey, 1850; Squier and Davis, 1848; Squier, 1850; Lapham, 1855; Haven, 1856). Such effective use of the publishing resources of the federal government put Henry in position to shape the structures of knowledge and practice in the nascent discipline (see Hinsley, 1981; Henry, 1996).

The Smithsonian's program of professionalization, however, was entirely reliant on a network of antiquarian collaborators. The Institution had no professional staff, and a limited budget to support research. Most of the *Contributions to Knowledge* studies arrived as substantive manuscripts, sent by antiquarian entrepreneurs from states where archaeological sites were being exposed by settlement, land clearance, and cadastral surveys. Shorter communications from interested parties also traveled along the network. There was little pattern to these reports, or consistency in their content. Artifacts were sent as well: managing this flow of information became a major problem. When naturalist Spencer Baird became Assistant Secretary he took personal interest in the collections (see Henson, 2000), but keeping track remained difficult.

The antiquarian communities of the United States had a generally positive response to the Smithsonian's collection of information about the indigenous past. The Institution's imprimatur was a mark of status and a source of potential support for local initiatives. Henry occasionally funded modest fieldwork and collecting projects, and such investments were eagerly sought. For instance, in February 1854 the vice-president of the New Orleans Academy of Sciences paid a call on Henry and Baird while visiting Washington: his report described the meeting as focused on shared collecting opportunities, as well as potential 'exchanges' that would 'make a very handsome nucleus for a museum for the Academy' (New Orleans Academy of Sciences, 1854: 62).

There was, however, inherent tension in the relationship between the Smithsonian and others in the antiquarian networks. As local institutions and scholarly 'circles' matured throughout the American south and midwest, tension between national sanction and local achievement became more common. Collectors or agents working in support of Smithsonian projects were seen as rivals to locally based entrepreneurs.

As the museum functions of the Smithsonian expanded after the Civil War, tension over the antiquarian capital represented by artifacts became particularly evident.

It was also anticipated that Smithsonian would 'synthesize' dispersed knowledge about American archaeology, both through the centralization of information and through quick publication. The need for such a catalog, map, or synthesis was widely discussed. 'It is to be hoped that the honored Secretary...' wrote John Wells Foster, 'will bring out an illustrated catalogue of American Antiquities, not restricted to Smithsonian collections, but embracing those of individuals throughout the United States' (1873: vi).

Henry, however, encouraged primary research rather than synthesis, and used the Smithsonian's scarce resources accordingly. '[I]t strikes me,' he wrote to one of his correspondents, Charles Rau, 'that all that is required at present, in the way of publication, is sketches of progress, suggestions of hypotheses to direct lines of research, and instructions as to the method of making explorations, and the preservation of relics, etc.' (Henry to Rau, June 6, 1868.)[1] Such 'sketches of progress,' however, failed to satisfy Henry's antiquarian constituents, who continued to send reports piecemeal to Washington. In the absence of a common strategy or a consensus about archaeological methods, correspondents continued to act according to their individualized preferences. Thus over time dissonance increased within the network, challenging the nascent structures of the profession.

American archaeology at the Centennial Exposition

The engagements and tensions of American antiquarian networks came into full view in the context of the 1876 Centennial Exposition, held in Philadelphia. Leaders of the Smithsonian clearly saw the event as an opportunity to showcase their vision for American scholarship, endeavors that included archaeology. Baird noted that 'in view of the very great interest in subjects of this character, it was determined [to exert] special effort to render the display exhaustive and complete' (1876: 67). Some formal collecting was sponsored, and items borrowed from other institutions (*cf.* Putnam, 1876: 7; Powers, 1877) (see Figure 2.1). But the Smithsonian's network – along with those of other federal agencies, such as the Department of the Interior – was essential to this effort. Collections were received, including

> many thousands of stone implements of every kind and character. In some instances gentlemen who were not willing to present their collections permanently, have consented to lend them for the Exhibition, and

(A.) No. 7.

International Exhibition, 1876.

BOARD ON BEHALF OF UNITED STATES EXECUTIVE DEPARTMENTS.

NATIONAL MUSEUM, SMITHSONIAN INSTITUTION.

COLLECTION TO ILLUSTRATE THE

MINERAL RESOURCES OF THE UNITED STATES.

Invoice of Specimens Collected at Cairo Ill.

By J. C. Zimmer Esqr.

NO.	DESCRIPTION.	LOCALITY.
1 – 7	Bottles two with images	Mississippi Co Missouri
7 – 14	Cups three ornamented	" "
14 – 21	Pestles, Missouri granite & blue stone	" "
21 – 25	Deep Dishes, Urns and Plates	Tennessee River
25 – 33	Hatchets, granite, blue and green stone	" "
33 – 40	Fleshers, Chert, granite, green and blue stone	" "
41	Sinker green stone,	" "
42	Gorget " "	" "
43	Rimmer " "	" "
44	Acorn marble	" "
45	Shovel chert	" "
46	Hoe "	" "
46 – 49	Images	Mis. Co Mo.
49 – 53	Balls blue stone	Tenn. River
53 – 113	Spear Points colored chert various	Ohio

2.1 Invoice of specimens obtained at Cairo, Illinois, from J.C. Zimmer, Esq., on behalf of the Smithsonian Institution, for the Centennial Exposition of 1876. CLE. Smithsonian Institution Archives, Image # SIA 2011-0783.

> these will be carefully kept separate and returned at its close (Baird, 1876: 69).

The Smithsonian's representative in Philadelphia was Henry's long-time correspondent Charles Rau, whose career reflects both the vicissitudes of local antiquarians in the United States and the new opportunities of the 1870s. Rau left Germany in 1848, pursuing a teaching career first in rural Illinois then in New York while keeping up correspondence on antiquities, amassing a considerable collection in the process (Kelly, 2002). His employment by the Smithsonian in May 1875 came after eighteen years of labor, and Rau was motivated to make the exposition a success (Henry to Rau, May 10, 1875).[2] The annual report for the 1876 year lists materials received from sixty-six different donors, only a few of whom were formally employed by the Institution (Smithsonian, 1877: 84–104). Most of this material would be added to the national collections after the Exposition ended, creating new opportunities for display and – potentially – scholarship.

The Smithsonian's determination to make the Centennial Exposition a showplace for archaeology was shared by local antiquarian societies across the country. An International Archaeological Congress held in September demonstrated their numbers and engagement (*cf.* Stocking, 1976). Exhibits featuring antiquities of all kinds were presented to the public. The display mounted by the Ohio State Archaeological Society, for example, occupied sixteen cases and included artifacts from forty-five different private collections. Described as a 'positively splendid collection' in Frank Leslie's guidebook (1876: 216), the Ohio exhibit demonstrated the potential for regional networks conducting their own archaeological projects. At the center of the display was a map of the state illustrating the distribution of different types of site and of the indigenous population. 'Much is to be added to this map to perfect it,' the organisers remarked. '...This is a work which must of necessity be done by the State, or it will never be thoroughly accomplished' (Read and Whittlesey, 1877: 82).

The upsurge in local archaeology on display at the Centennial Exposition also reflected changing attitudes toward the Smithsonian's antiquarian network. Protest over the flow of collections to Washington was widespread. In particular, the Institution's failure to coordinate and synthesize the influx of information was seen as hindering the collective effort. Roeliff Brinkerhoff, the president of the Ohio Archaeological Society, expressed this concern publicly during the Archaeological Congress. 'We want,' he announced in his opening speech, 'some means of communication so that each may be kept advised of progress made, and whenever any point is gained all may know of it, and no further efforts shall be wasted in determining what has already been determined' (*The Ohio Liberal* 4(22), Mansfield, Ohio, September 13, 1876).

As an alternative to Smithsonian centralization, Brinkerhoff and his colleagues at the Ohio Society launched a new, decentralized network, designated the 'American Anthropological Association' (Stocking, 1976).[3] Focused on local societies and a more open information exchange, the organization would effectively bypass the Smithsonian and similar 'professionalizing' efforts under way by other, nascent national institutions.

From the perspective of Henry and Baird, the Smithsonian's 'success' at Philadelphia was immediately threatened by the divergent aims of their colleagues in the Archaeological Congress. Faced with the potential unraveling of their network, the two abruptly changed policy and launched an effort to synthesize the results of American archaeology. The Smithsonian staff remained modest, but Henry was committed to reinforcing the Institution's central role in antiquarian networks. Compiling archaeological information would take advantage of the public's interest and allow alternative models to be bypassed. There was still no clear sense of how such antiquarian knowledge might be processed into a coherent whole, but that problem seemed less significant than threats to the Smithsonian's central role in American antiquarian networks.

Circular 316

Rather than develop a new approach to collecting information about American antiquities, Henry opted for a more traditional strategy: the circular. Antiquarian societies in the United States had used circulars since at least 1797, when the American Philosophical Society in Philadelphia had requested that its correspondents send 'accurate plans, drawings and descriptions... of the ancient Fortifications, Tumuli, and other Indian works of art: ascertaining the materials composing them, their contents, the purposes for which they were probably designed, &c.' (American Philosophical Society, 1799: xxxvii; Smith, 1996: 7). Another circular was sent out by the War Department in the 1820s to elicit first-hand accounts about Native American languages. Its designer, Albert Gallatin, argued that the resulting information would be an important scholarly resource (Bieder, 1986: 29).

The antiquities query sent by the Smithsonian in the spring of 1878 was designated Circular 316, 'In Reference to American Archaeology.' Otis Mason, a professor at Columbian College who had been assisting Henry and Baird with collections projects, prepared the fifteen-page document, which solicited information on twenty-one different categories of antiquities, including sites, locations, and collections. The section on 'Mounds and Earthworks' requested that a map be included, 'however rude it may be… A topographical survey is most desirable, when it is convenient'.[4]

Mason tallied approximately 216 responses to Circular 316 received through 1880 (Mason, 1880): associated files in the Smithsonian archives include material from approximately 499 different correspondents dating between May 1876 and December 1879. The flow of correspondence was greatest in the period immediately after the circular was sent, with at least eighty-five letters arriving in May 1878 alone. Some respondents to the circular were local experts who had been part of the Smithsonian's network for years. These include the Rev. William M. Beauchamp, of Baldwinsville, New York, who would go on to publish widely on Iroquois topics, and Matthew Canfield Read, one of the organizers of the Ohio exhibit at the Centennial Exposition (Read to Baird, September 5, 1878).[5] Another, Charles Metz, would shortly become a collaborator of the Peabody Museum of Archaeology and Ethnology at Harvard (Burns, 2008).

The vast majority of responses, however, were sent by people who encountered archaeological remains in the course of their daily lives. Most preferred to make general comments about their experiences rather than comply with the requested format. Many correspondents provided information about individuals in their counties or towns who either took a general interest in antiquities or had collections of their own. In many cases the letters were accompanied by sketch maps of sites and, in particular, drawings of artifacts. Many of these were essentially marginalia, intended to convey a general sense for the material. In a few instances these were well executed and conveyed a clearer understanding of the artifacts concerned (see, e.g. Figure 2.2). But a more typical example is a letter from Dr Moses Quinn MD, of Dalton, Georgia, indicating that he had seen 'aboriginal remains scattered all over the country,' but including no details (Quinn to Henry, 1 June 1878).[6]

Longer-term engagement with correspondents was rare. Only twenty-nine individuals wrote three or more times, but some of these correspondents were deeply interested in the project. One of these was Samuel B. Evans, of Ottumwa, Iowa. Evans was a journalist, writing for the *Ottumwa Democrat,* a platform that he had used for articles on the 'mound builders.' 'I have given the mounds considerable attention,' he wrote, 'and have explored many of them in this county and have my mind on a few more which I wish to go through before giving an elaborate report.' A month later he wrote again, having

> just returned from Van Buren County in this state where in company with Judge Sloan and D.C. Beamon, Esq., of that Co. I explored four

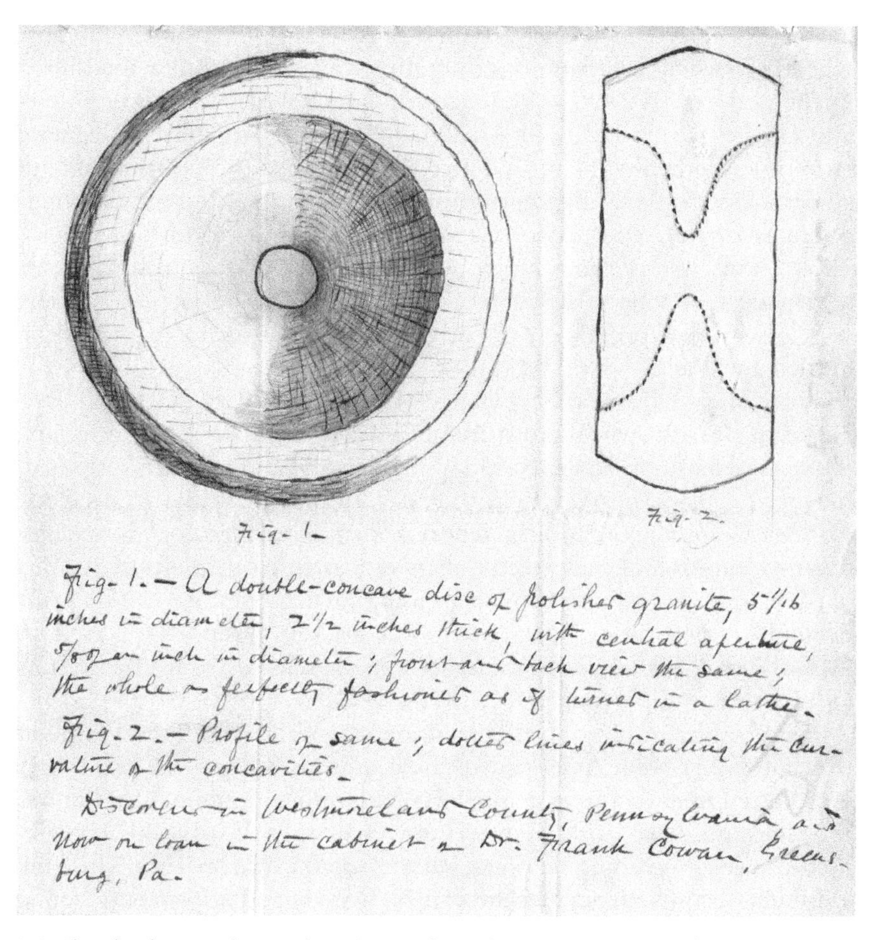

2.2 Sketch of an artifact in the cabinet of Frank Cowan, Westmoreland County, Pennsylvania. CLE. Smithsonian Institution Archives, Image # SIA 2011-0784. Copyright © Smithsonian Institution Archives, Washington, DC–CLE. All rights reserved and permission to use the figure must be obtained from the copyright holder.

> mounds, and succeeded in securing a skull of a mound-builder in a pretty good state of preservation.

Some material from this effort was subsequently shared, and an undated response from Rau expressed thanks 'for plaster casts. Report will be utilised in the work on mounds to be written by Professor Mason' (Rau n.d. [response to Evans]).[7]

Many of the respondents, like Evans, were of the 'educated' classes in their local community. Doctors, teachers, and attorneys are on the list, as are pharmacists, postmasters, bankers, pastors, judges, civil engineers, and one jeweler – Charles Artes, of Evansville, Indiana, whose extensive collection of antiquities was widely commented upon in the era (e.g. W.M. Locke to Baird, January 8, 1879.[8] See Wilcox and Hinsley, 2003.) One of Artes' neighbors, F. Stinson, was general Secretary of the Evanston Young Men's Christian Association, but spent his spare time 'digging into mounds & graves, taking drawings & items of the same, & of earthworks' (Stinson to 'Sirs,' May 10, 1878).[9] Fully ten of the correspondents were the editors of small-town newspapers, serving as nodes of contact for their community. Farmers were also well represented, often reporting about artifacts plowed up in their own fields.

Very few of the correspondents were women. Three – Sarah Jane Foster of Beardstown, Illinois; Annie L. Peyton of The Plains, Virginia; and Cara Chase of Wells, Nevada – made substantive contributions to the effort. Foster, in particular, took a deep interest in the archaeology of Cass County, preparing a report and maintaining correspondence over the duration of the project. There is also ample evidence of women participating at least indirectly in antiquarian activities. F.M. Witter, for example, of Muscatine, Iowa, put in a request for a report to be sent to 'Miss Edith Winslow,' and there are similar references in other letters (Witter to Baird, November 16, 1878).[10]

Letters were received from forty-three states and territories. Ohio – at the center of popular American antiquarianism since early in the century – produced thirty responses. Similarly high numbers came from Illinois, Indiana, and New York. Representatives of local institutions contributed as well, including Thomas Rhodes of the Akron City Museum, who sent a sample from a cache of 190 'leaf-shaped implements' found 'under an old tamarack stump' (Rhodes to Baird, January 10, 1879) (see Figure 2.3).[11]

Responses from the western states and territories were few, most contributed by military officers, members of exploring parties, and government officials. Some came indirectly – written to officials within their own hierarchies, who then passed them along. A captain of the 8th Infantry stationed at Camp Verde, Arizona, sent a box of artifacts dug up in nearby ruins, while an agent of the General Land Office agitated for preservation of the ruins at Casa Grande (G.M. Brayton to Officer in Charge, Army Medical Museum, February 23, 1878; Charles D. Poston to Baird, November 30, 1878).[12]

Evidence for local antiquarian networks is evident in many of the Circular 316 letters. The Iowa networks were particularly robust. The Davenport Academy was a major contributor to archaeological knowledge during this period, and several of its members corresponded with the Smithsonian (e.g. W.H. Pratt to Baird, February 17, 1877;

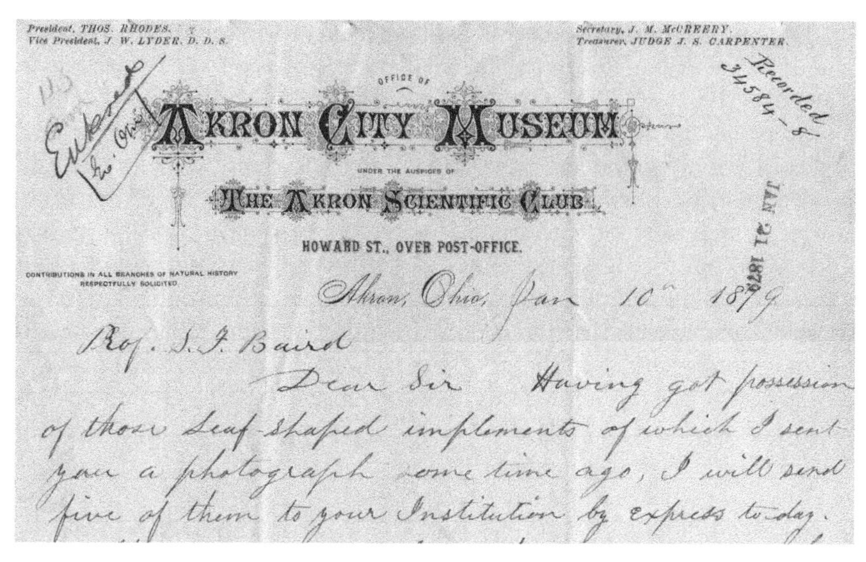

President, THOS. RHODES.
Vice President, J. W. LYDER, D. D. S.

Secretary, J. M. McCREERY.
Treasurer, JUDGE J. S. CARPENTER.

OFFICE OF
AKRON CITY MUSEUM
UNDER THE AUSPICES OF
THE AKRON SCIENTIFIC CLUB
HOWARD ST., OVER POST-OFFICE.

CONTRIBUTIONS IN ALL BRANCHES OF NATURAL HISTORY RESPECTFULLY SOLICITED.

Recorded 34584–8

JAN 21 1879

Akron, Ohio, Jan 10th 1879

Prof. S. F. Baird

Dear Sir Having got possession of those Leaf-shaped implements of which I sent you a photograph some time ago, I will send five of them to your Institution by express to-day.

2.3 Akron City Museum letterhead. CLE. Smithsonian Institution Archives, Image # SIA 2011-0790. Copyright © Smithsonian Institution Archives, Washington, DC. All rights reserved and permission to use the figure must be obtained from the copyright holder.

R.J. Farmingham to Baird, February 24, 1877; J. Duncan Putnam to Baird, November 8, 1878.[13] *Cf.* Goldstein, 2008.) It was not the only organization in the state devoting attention to antiquarian matters: F.M. Witter, of Muscatine, described the local 'academy of sciences' as well as efforts by the city to hire an archaeologist (Witter to Baird, November 16, 30, 1878).[14] Seth Dean, a county surveyor based in Glenwood, recommended the Council Bluffs missionary Samuel Allis, who returned the favor: the two also exchanged artifacts and consulted on questions from the circular (Dean to Henry, April 10, 1878; Allis to Henry, May 22, 1878).[15]

The underlying tensions within the antiquarian community of the late-nineteenth-century USA is evident throughout the Circular 316 collection. The desirability for coordination is frequently expressed, particularly through the circulation of information about the substance and practice of archaeology. Joel Barber, county surveyor in Lancaster, Wisconsin, wrote 'I rejoice at the effort you are making. We needed such a pamphlet to teach us what and how to report' (Barber to 'Sir,' April 13, 1878).[16] Dozens requested copies of the Institution's reports. Smithsonian employment or funding was widely desired, since otherwise few had the resources to pursue such interests. 'We are all more or less poor, and can depend only on individual interest in the work,' noted one correspondent (James Pomeroy to Secretary, July 30, 1878).[17]

The task of synthesizing this amalgam of information fell to Mason and Rau, who had become a permanent member of the Smithsonian staff after Philadelphia. Neither was adept at the particular demands of the project, and it is clear that they despaired of the task. Mason's experience with American archaeology 'on the ground' was limited, and he demonstrated little understanding of the challenges faced by those collecting the sort of information desired. Rau's expertise pertained to artifacts, particularly stone tools, and he took less interest in other categories of information. From his perspective most of the material received was useless for the planned synthesis. Rau's curt assessment of an artifact sent by William Taylor, a farmer from Scarborough, Tennessee, preserved in the margin of the letter, was typical. 'No answer required. This spec. amounts to nothing' (Taylor to Baird, November 10, 1878).[18]

Networks and synthesis

The deepest irony of the Circular 316 project was that it failed almost before it began. On May 13, 1878 – shortly after the circular was released – Joseph Henry died in the Smithsonian Castle. He was succeeded by Spencer Baird, who had a different agenda. Although Baird supported archaeology, particularly the acquiring of materials for the museum, he seems to have been less concerned about establishing institutional priority. Unlike his predecessor Baird was in close personal contact with many in the national network of natural history collectors, and seems to have had a better grasp of the strengths and weaknesses of antiquarian 'crowdsourcing.'

A more fundamental shift that rendered the Circular 316 information less useful for its original purpose was the creation, in 1879, of the first 'professional' anthropological organization in the United States, the Bureau of Ethnology (later, the Bureau of American Ethnology). Directed by John Wesley Powell, the dean of the federal survey teams that had documented the west, the bureau was placed under Baird's nominal authority but was run as an essentially independent body. Powell had corresponded with Henry as part of the Circular 316 project, but once in a position of authority rejected the populist, networking strategy it represented. Instead, he assembled a research team made up of war veterans and former survey party members, who ran their own field operations and ethnographic studies under his overall direction. In such a scheme, local knowledge and networks had limited value.

For a period of time Mason and Rau continued their efforts, and are mentioned approvingly in various annual reports over the next few years.

2.4 S.T. Walker, 'Mound at Bullfrog [Florida]'. CLE. Smithsonian Institution Archives, Image # SIA 2011-0794. Copyright © Smithsonian Institution Archives, Washington, DC. All rights reserved and permission to use the figure must be obtained from the copyright holder.

Both men had other projects to attend to, however, and eventually the Circular 316 letters were filed away. A selection of the more thorough accounts provided was published in the print outlets of the Smithsonian, including the Annual Reports. The work of Evans and his colleagues on Iowa mounds, for example, appeared in 1880. The unsystematic nature of this material, however, meant that it was difficult to review in any syncretic way, and it remains almost impossible to cite.

As historian Curtis Hinsley has noted, '... the circulars provided... insufficient control over personal idiosyncracy,' and it is true that their value for generating archaeological information is slight (Hinsley 1981: 152). The great value of the Circular 316 project for historical scholarship, however, is the collective image the correspondence depicts of archaeological activity by the American public, particularly in rural areas. Through text, sketches, maps, and photographs, an image created of a population deeply involved with the material past.

Notes

1 Charles Rau Papers (CRP), Record Unit 7070. Folder 2. Joseph Henry, 1866–70. Smithsonian Institution Archives, Washington, DC.
2 CRP, Folder 2. Joseph Henry, 1871–6.

3 As Stocking (1976) notes, there is no apparent connection between this organization and the modern American Anthropological Association, founded in 1902.
4 Circular 316. Collected Letters on Ethnology (CLE), Record Unit 58. Smithsonian Institution Archives, Washington, DC.
5 CLE.
6 CLE.
7 CLE.
8 CLE.
9 CLE.
10 CLE.
11 CLE.
12 CLE.
13 CLE.
14 CLE.
15 CLE.
16 CLE.
17 CLE.
18 CLE.

3

'More for beauty than for rarity': the key role of the Italian antiquarian market in the inception of American Classical art collections during the late-nineteenth century

Francesca de Tomasi

Introduction

> It is very surprising that there has been a buyer of such vases unless we assume that they were destined for some American museum; since everybody knows that the Americans, without any particular knowledge of art history and without leaving their country of origin, buy art on commission. They trust the archaeological knowledge of the people they appoint for the purchases. If this attitude were true it should be deplored. That is, those who should be in charge of the study and the conservation of the national cultural heritage procure its export abroad instead. Neither is it commendable to sell foreigners (even if Americans) mediocre objects as if they were works of art or monuments with a true value or archaeological interest (Chigi to Fiorelli, December, 1889).[1]

Although this excerpt is not from an official document but from a private letter addressed to the head of the Directorate General of Antiquities and Fine Arts by an employee of the office, the quotation above makes it clear that the idea that Americans were not able to understand and evaluate Classical art was taken for granted in late-nineteenth-century Italy and even pervaded state institutions. The statement shows how both the artistic taste and the connoisseurship of the American collectors were underestimated by Roman scholars when the collectors were first encountered at the end of the nineteenth century. Clearly, this superficial statement did not take into account the real damage that the growing interest in antiquities in the United States could have caused – and eventually did cause – in Italy.

The purpose of this chapter is to describe a particular historical period, which runs from the late 1880s to the first decade of the twentieth century, when American collectors and museums began to express interest in purchasing antiquities from the Mediterranean area and particularly from Italy. During this first period, though, the Americans timidly approached the Italian market, starting to purchase small objects via occasional intermediaries. Two cases, reconstructed using archival sources, will show how Latour's actor-network theory (1987, 2005) can be applied to explain the dynamics of the antiquarian trade between Italy and the United States in that particular time frame. The archaeological objects, their true or presumed place of discovery, the scholars who studied them and the art dealers and collectors who sold and bought them can be interpreted as actors/actants able to produce and spread scientific knowledge.

The geographies of knowledge theory (Livingstone, 2003) is also useful to highlight the existence of a strict relationship between the 'sites' where knowledge was created, the subjects that were allowed to access those 'sites' and the way that knowledge was spread. Livingstone writes:

> Just how knowledge embedded in a particular location moves from its point of origin to general circulation, and thereby transcends locale, is an inherently spatial question and introduces a crucial dynamic to the geography of science. Rather than being understood simply as an inevitable consequence of a uniformly constant nature, the universality of science is the consequence of a variety of practices that have had to be put in place to guarantee reliable transmission (Livingstone, 2003: 181).

In the two cases examined, networks of scholars, dealers and collectors played a key role in promoting increasing purchases of antiquities outside Italy.

At the end of the nineteenth century, many Italian scholars of Classical antiquity believed they held the 'gold standard' in their field of study. A few years earlier, the Director-General of Antiquities and Fine Arts Giuseppe Fiorelli had written in an official report on the service he directed:

> the foreigners themselves, who contend with us for possession of our objects, need an official seal that only the Government authority can give, which is fundamental to assess the scientific value of those objects (Fiorelli, 1885).[2]

Accordingly, Fiorelli continues, foreign museums could increase the value of their collections only by buying objects which had been studied in Italy. In this way already well documented archaeological finds would not be exported abroad as 'refugees', but as 'settlers' of Italian culture.

The 'settlers', as Fiorelli refers to them, are akin to the concept of 'immutable mobiles' that Latour developed in 1987. They are a form of knowledge in movement, they can cover distances and they can make science accessible far from its 'site' of origin (Latour, 1987, 2005).

In late-nineteenth-century Rome the exporting of archaeological finds and of antiquities from private collections was already a widespread phenomenon. Many collectors, art dealers and intermediaries gained great benefits and personal profit. The market was very fluid and it benefited from the lack of strict legislation. After unification of the Kingdom of Italy in 1861, it took many years before a law was passed to regulate the circulation of antiquarian objects. The main difficulty in regulating the export of antiquities was the Italian Parliament's clear intention not to limit private property in any way. Anyone who owned an archaeological object or a piece of art could dispose of it, like any other goods. Accordingly, the antiquarian market could take advantage of this situation at least until 1902 when eventually a law was passed (Law of 12 June 1902, n. 185), but it was only in 1909 that the law became coherent and effective (Law of 20 June 1909, n. 364) (Mattaliano, 1975: 3–89; Bencivenni, Dalla Negra, Grifoni, 1987, 1992; Barnabei and Delpino, 1991). As a consequence it very often happened that the network of the experts who could establish the importance – and the value – of an archaeological piece coincided with that of the state officials in charge of protecting cultural heritage. No scholar, archaeologist or even state official had not been called upon at least once to give an opinion on a purchase, to write a report to support the granting of an export permit or to provide an estimate of the market value of an artwork. Consequently, they often crossed the boundaries between archaeology and antiquarianism, conservation and collecting, legal and illicit. As a result, often scholars who were in contact with the unofficial channels which operated outside the law were the first to be informed of new finds or of the sale of pieces belonging to private collections.

Such was the case with two of the most famous and successful archaeologists of the time: Wolfgang Helbig and Rodolfo Lanciani. Both Lanciani and Helbig were active in a period when the possibility of creating and increasing knowledge about the ancient city of Rome was higher than it had been for centuries. The transformation of the Papal city into the new capital of the Kingdom made necessary a series of urban projects to build roads, government buildings and new districts. The high number of construction sites was revealing substantial portions of the ancient city underneath the modern one.

Lanciani's work has the considerable value of documenting what was being uncovered beneath Rome in those years while Helbig's work is of fundamental importance for showing what collections of antiquities existed in Rome at the beginning of the twentieth century (Lanciani, 1893–1901; Helbig, 1891). Though each of them had a dynamic personality and was highly regarded in his field of study, they operated in a slightly different way in dealing with the antiquarian market and its actors.

Wolfgang Helbig (1839–1915)

From 1865 Helbig was the Second Secretary of the German Archaeological Institute in Rome. In 1866 he had married Nadine, a Russian princess, and the social contacts he gained by the marriage made his house an international cultural salon (Helbig, 1927; Blanck, 2004: 670–73; Örma and Sandberg, 2011; Moltesen, 2012: 35–49). After 1887, when he left his role at the Institute, Helbig became close to Carl Jacobsen (1842–1914), 'the brewer' from Copenhagen, and assumed the role of Jacobsen's purchasing agent in Italy with a salary of 5,000 French francs a year for twenty-five years. This partnership allowed Jacobsen to purchase around 900 archaeological objects from Italy – some of them of considerable value – to be exhibited in the museum he planned to donate to his city: the Ny Carlsberg Glyptotek (Moltesen, 2012).

In the Roman intellectual salons – such as the Helbigs' house – antiquarianism was still a favourite topic of interest and represented a form of social amusement. Members of the international cultural aristocracy would gather to discuss new discoveries, to show off their collections and to establish networks that would allow them enrich those collections (Pollak, 1994). These circles – which continued the tradition of Classical studies as erudition – hosted scholars, collectors and art dealers in a mutual exchange of objects, information and scientific recognition that was often confused with social prestige.

Among Helbig's most trusted friends were antiquarians such as Francesco Martinetti (1833–95) and collectors such as Michal Tyskiewicz (1828–97), as well as many nobles (such as the Sciarra, Barberini, Odescalchi, Boncompagni Ludovisi, Borghese, Orsini and Giustiniani families) who had fallen on hard times and wished to sell their collections on the antiquarian market (Tyskiewicz, 1895, 1896a, 1896b, 1897a, 1897b; Jandolo, 1935: 37–41; Barnabei and Delpino, 1991: 162–5; Pollak, 1994: 134, 221; Molinari, 1994; Moltesen, 2012: 161–88). Helbig played, without a doubt, a leading role in this antiquarian trade. Not only Jacobsen and many nobles in financial difficulties, but also Americans – who were accessing the antiquities market for the first time – very soon realised that Helbig was the person they should consult. In 1891 he wrote to Jacobsen: 'Therefore I don't fear the European museums so much as the Americans who have only scorn

3.1 Bronze statue of a *camillus* donated by H.G. Marquand to the Metropolitan Museum in 1897 (MMA 97.22.25). Public domain image.

for entails and the sort. I am, however, assured that they do not yet show much striking power' (Moltesen, 2012: 161). Thus, Edward Perry Warren (1860–1928), intermediary for the Boston Museum of Fine Arts from 1885, and his friend John Marshall (1862–1928), became clients of Helbig (Burdett and Goddard, 1941; Green, 1989; Sox, 1991; Murley, 2012). In addition, Henry Gurdon Marquand (1819–1902), the President of the New York Metropolitan Museum's Board of Trustees, acquired antiquities from the Roman market between 1887 and 1896 through the mediation of Helbig (Del Collo, 2011). Marquand was a railway tycoon and became a trustee of the Metropolitan Museum in 1871. In 1889 he was appointed the Museum's second president. In its archives there are still traces of the relationship between Marquand and Helbig.[3] Arthur Lincoln Frothingham (1852–1923) who, between 1894 and 1896, worked at the American School of Classical Studies in Rome and acted as purchasing agent for various American museums, was also involved in the sale (Lavin Aronberg, 1983: 14–18; De Puma, 1996: 471). The purchases consisted of a series of small and medium-sized bronze objects which Marquand personally paid for and later donated to the Museum.[4] The most important purchases he made were a bronze statue of a *camillus* (see Figure 3.1)[5] – at the time considered to be of the emperor Geta – and a bronze statuette of Cybele on a cart drawn by lions (see Figure 3.2).[6] On 22 October 1887, Helbig sent a letter to advise that he had received the 20,000 French francs Marquand had sent to him and had used them to buy the head of the statue of Geta. The body of the statue was still in the hands of the owners with whom Helbig could not have made a legal contract because the whole sale was illegal. Helbig wrote to Marquand:

> In case of a breach of contract *[sic]* I could hardly lodge a complaint as my own complicity as the affair would involve me in legal difficulties. Besides, in such critical cases as this I have found the Italians more honourable than under normal conditions (Helbig to Marquand, 22 October 1887).[7]

The *camillus* was the more expensive purchase, costing 60,000 *lire* plus 2,500 *lire* for export duty and 10.25 *lire* for transportation. Once acquired, the bronze required restoration. For this task, Helbig proposed a friend from Rome as the only person both trustworthy and capable of keeping the deal secret. This man was Martinetti. In 1888 he took charge of the restoration and the packaging of the statue. Arthur Frothingham saw the statue in June 1889 and found it 'a very fine work of art', and he informed Marquand that it would be sent the week after 'with the diplomatist' *[sic]* luggage, in a good shape.'[8]

It took longer to bring the statuette of Cybele to New York. Problems arose with the owners, possibly over the export licence. On 2 May 1896 Helbig wrote:

> I thank you for the 30,000 fcs, which brought in exchange 32,200 *lire*. The additional exchange I have not been able, unfortunately, to pay Mr Frothingham as I had expressly contracted with the owners of the

3.2 Bronze statuette of Cybele on a cart drawn by lions, donated by H.G. Marquand to the Metropolitan Museum in 1897 (MMA 97.22.24) Public domain image.

> group at a price of 30,000 francs plus 6,393.25 *lire* (the balance on hand), and this stipulation was authorised by your letter of December 11th, 1895. You get out of the transaction very well indeed (Helbig to Marquand, 2 May 1896).

The chariot was then shipped to Paris in order to be restored. Helbig continued:

> I am naturally extremely anxious that you should inform no one that I have had a part in this purchase, and it is also advisable to urge Mr Andre not to show the group to anybody in Paris. It is to be hoped that Mr Andrew *[sic]* will restore the group of Cybele as well as he has done the silver pieces from Boscoreale, whose restoration appears to have been a masterpiece (Helbig to Marquand, 2 May 1896).[9]

In May 1897, after the chariot arrived, Marquand donated it to the Museum together with some other objects he had bought: twenty-five Greek, Roman and Etruscan bronzes, statuettes, busts and mirrors. Among the bronzes donated to the Metropolitan Museum are the following items:[10] a statuette of an *ephebe* resembling the Polykleitos *Doryphoros* and *Diadoumenos*;[11] a fragment of a moulding from the Pantheon;[12] the head of a grotesque from the Tiber;[13] a statuette of Aphrodite with apple from Santa Maria Capua Vetere;[14] two Etruscan mirrors;[15] a helmeted bust of Minerva;[16] and a statuette of Jupiter Capitolinus.[17] Other objects remained in his private collection which was sent to auction in 1903 after his death (Kirby, 1903).[18] In the *Boston Evening Transcript* of 30 January 1903 an anonymous journalist, describing Marquand's activity, wrote that the collector had bought objects 'more for beauty than for rarity' (anon., 1903).

Helbig expressed the wish to remain anonymous several times in his letters. This is the reason why his name never appears in the catalogues published by the Metropolitan Museum. It may be wondered why Helbig wished to keep his purchasing activity secret since it was probably public knowledge – judging by the number of clients and contacts he had. Carl Jacobsen even gave Helbig's name to the Etruscan Gallery of the Ny Carlsberg to pay homage to his friend (Poulsen, 1927). Was it perhaps that he did not want to enrage the Italian authorities? Or that he did not wish to be associated with small, occasional and worthless purchases? The question remains unanswered. However, we can hypothesise. We can take as an example the well known case of the *Fibula prenestina*, a gold *fibula* with an inscription that seems to be one of the first written attestations of the Latin language. Thanks to his friendship with the art dealer Francesco Martinetti, Helbig had the opportunity to be the first to see the object. He was also the first to publish an article on the *fibula*, strongly linking his name to it (Helbig, 1887). In recent years, its dubious origin – place of discovery unknown – led the *fibula* to be considered a fake created by Martinetti and validated by Helbig (Guarducci, 1980, 1984). Recent studies, though, have demonstrated the authenticity both of the object and the inscription (Franchi De Bellis, 2011; De Simone, 2011; di Gennaro et al., 2015). The *fibula* affair shows that the dubious or uncertain origin of the archaeological material did not discourage Helbig from publishing his article. It seems more likely then that Helbig did not want his name to appear in the catalogues of the Marquand collection in order not to reveal his network of contacts, especially in a rising market such as the American antiquarian market. It was probably more important to him to be officially part of an Italian/European network and to occasionally work for American collectors and museums only as a source of income.

Rodolfo Lanciani (1845–1929)

During the same period another episode occurred which involved the renowned archaeologist Rodolfo Lanciani, who worked from an early age for the Italian government, for Rome's city council and for La Sapienza University in Rome, where he was Professor of Ancient Topography between 1894 and 1922 (Palombi, 2006).[19] Lanciani had always particularly admired the Anglo-Saxon world and in 1886–87 he was invited by several American universities – including Harvard, Princeton, Pennsylvania and Columbia Universities – to give a series of lectures, which were met with huge success (Palombi, 2006: 113–22). It was probably on these occasions that he met the directors of the Chicago and Boston museums who asked him to contribute to the growth of the museums' collections. In the 1950 Boston museum catalogue can be read 'Mr Robinson immediately began to urge upon the trustees the importance of acquiring original works of art, especially sculpture and vases, and in 1888, with the help of Professor Rodolfo Lanciani, whose interest in the Museum had been aroused when he came to Boston to lecture in 1887, a number of marbles, heads and portrait busts, as well as selected terracottas, bronzes, vases and coins from Italy were purchased' (Chase, 1950: 1).

It was his agreeing to assist US cultural institutions that cost Lanciani his reputation. In 1889, the archaeologist was accused by the Italian Minister of Public Education of having acted as intermediary for the sale of numerous artefacts to the United States. The event represented a significant setback in Lanciani's professional life; at the peak of his career he was forced out of his professional positions, except for his teaching post at La Sapienza University (Palombi, 2006: 122–47). Ironically, the first accusations were made by the Director of the New York Metropolitan Museum: Count Luigi Palma di Cesnola (1832–1904) – an Italian who certainly had another motive than the protection of the antiquities in his country of origin – reported the practice of antiquities trafficking (Roversi, 1898; McFadden, 1971; Moncasoli Tribone and Preacco, 2004; Damilano, 2014). Subsequently, the Ministry of Public Education launched an investigation during which depositions were collected from Lanciani's colleagues confirming his connection to the antiquarian market (Barnabei and Delpino, 1991: 458–62, documents 6 and 8). Of the famous Lanciani Inquiry there is no trace in the official publications of the Ministry of Education. However, the documents were scrupulously safeguarded by Felice Barnabei, Deputy Director of Antiquities and Fine Arts and arch-enemy of Lanciani.[20]

What is certainly true is that from December 1887 Lanciani corresponded with Charles Loring and Edward Robinson, respectively the

President and the Director of the Boston Museum of Fine Arts, proposing works of art that he would acquire on their behalf from the antiquities market of Rome (Whitehill, 1970: 21–2; Dyson, 1998: 133).[21] Charles G. Loring (1828–1902) had been a general in the American Civil War and became the first President of the Boston Museum of Fine Arts. In 1885 Edward Robinson (1858–1931) was appointed Curator of Classical Art and in 1902 Director of the Boston Museum of Fine Arts. In 1905 he resigned and later that year he was hired by the New York Metropolitan Museum.

The relations between Lanciani and his fellow scholars were cordial; also, their families were involved in a reciprocal exchange of greetings and invitations, both to America and to Italy. Probably as a result of earlier agreements, the Boston Museum placed a sum of money at Lanciani's disposal with Rothschild's Italian branch. In January 1888, Lanciani complained to Loring about the immobility of the antiquarian market after the scandal at the building site for Vittorio Emanuele's monument (Coppola, 2009). Many coins found during the construction work were illegally sold by the workers and from that point the authorities became more suspicious. In his letters Lanciani focused on the market's dynamics:

> [B]etter to bargain with the producers: only these producers are becoming a myth! Since I came back two excavations only have been undertaken by private individuals: one in the Artemisium of Nemi by Signor Boccanera one in the Caere necropolis by the town clerk of Cerveteri. The Boccanera finds were *hors ligne*; and out of question for the sum you kindly put at my disposal. There was a statue of a man named Fundilius, a greek roman masterpiece of the augustan era *[sic]*; how I wished I could buy it for you! It has gone now to Denmark price paid 1400 dollars (Lanciani to Loring, 6 January 1888, original emphasis).[22]

He also complained about the difficulty in obtaining valuable objects:

> I firmly believe that, as far as the Italian antiquarian markets (the roman *[sic]* especially) are concerned, there are only two mines open for foreign Museums – namely – terracottas and busts. Statues, bronzes, inscriptions, mosaic works are out of question. Government and municipalities are extremely jealous of anything which may be suspected, right or wrong, to be of local interest: and strict orders have been given to the officers of the export bureau (Rosa, Tadolini and De Sanctis) to stop the migration abroad of first rate works and objects of local interest. The field within which limits I was allowed to act is exceedingly limited, still I hope to have succeded *[sic]* to your full satisfaction (Lanciani to Loring, 30 May 1888).[23]

Furthermore, Lanciani regularly advised his correspondents about the antiquities market and the direction the museum should follow in order to expand its collections:

> The experience I have gained in these last months tells me that the only archaeological objects and works of classic art that can be found on the Roman and Italian market, are terracotta, busts and coins. Would you believe that since October '87 only four inscriptions have been put out for sale? I have, of course, given up the idea of making a collection of them, from a chrono-paleographic point of view. My hope is that the authorities of the Boston Museum will decide on a very limited number of 'specialties' in future acquisitions. Would you not like the idea to get in the Museum the finest collection of busts in Europe and America, Italy excluded? or else the finest collection of terracottas? I believe it could be done in a short time (Lanciani to Loring, 16 February 1888).[24]

Despite the market crisis, Lanciani managed to buy several objects between 1888 and 1889.[25] In these instances, he made the purchases personally and, talking about some terracottas from Cerveteri,[26] he wrote:

> I have selected the specimens with the view that they should exhibit – in a small scale – all the characteristics of the great Cerveteri find. Gave 600 *lire* for them... The price is perhaps a little high but I had to contend against Helbig who was bidding for the Berlin Museum, and against Alberici who has become the most powerful monopoliser of antiques in Rome (Lanciani to Loring, 27 February 1888).[27]

Among the first purchases made in the winter of 1888, there can be found several busts,[28] and a series of coins.[29] In March 1888, Lanciani was able to buy further artefacts from Rome and its surroundings,[30] and another series of busts.[31] During the following winter Lanciani again bought numerous terracottas from the Tiber,[32] and a series of busts sold as items probably belonging to the Ludovisi collection.[33] The final shipment is dated to April 1889, shortly before the scandal broke, and it contained a series of ceramics from southern Italy.[34]

Furthermore, Cesnola accused Lanciani of also purchasing a number of vessels and artefacts which had featured in the 1889 Chicago Art Institute catalogue, such as Greek vases purchased by the Neapolitan judge Augusto Mele,[35] and objects from contemporary Roman excavations sold by the antiquarian Alberici, including busts, marble and terracotta items.[36] In fact, these items had been purchased by Charles Hutchinson (1854–1924) – President of the Museum – and by William French (1834–1914) – Director of the Art Institute – during a visit to Italy in 1888–89 (Alexander, 1994; De Puma, 1994; Hillard, 2010). In his travel notebook, French often mentioned Lanciani, who undoubtedly offered the two men valuable advice and useful addresses. Lanciani suggested the names of many art dealers (Marinangeli, Alberici, Innocenti and Fausti in particular) and French annotated the list in his notebook with the addresses of the shops; on 8 April 1889, he wrote a list of the objects purchased (French, 1889: 16–18). In truth Lanciani did have a friendly correspondence with Hutchinson at the Boston Museum, advising him on the procedures for requesting an export permit, packing and shipping.[37] Lanciani had received a sum of money to spend on behalf of the Chicago Art Institute which, however, he decided to send back in 1890 because of the problems that the investigation were causing him.

Meanwhile, by 12 March 1890 the report of the Ministerial Committee of Inquiry had been issued (Barnabei and Delpino, 1991: 472, document 16). The report focused on allegations against Lanciani, who in his capacity as a government official was accused of travelling to America to seek business clients, grant permits to excavate and request a subscription for the National Roman Museum; it noted that he held a monopoly on the excavations and discoveries in Rome. Up until then, Lanciani had not been able to defend himself. He did so in private with a letter to Fiorelli (Barnabei and Delpino, 1991: 464, document 10), and officially with a statement of defence (Barnabei and Delpino, 1991: 472–6, document 18),[38] that just preceded his request for exemption from his official post at the Ministry of Public Education on 30 December 1890. In both cases, Lanciani admitted that he had offered informal consultancy to the directors of Boston and Chicago museums, claiming he provided consultancy only for absolutely legal and traceable purchases.

However, the role played by Lanciani, as it emerges from the documents in Boston and Chicago, was not simply that of a consultant. The archaeologist was personally involved in the selection and purchase of objects and he knew how to act in the market and which dealers to approach. Not only was he aware of the bureaucratic procedures for obtaining export licences; he also obtained them for his clients. On 27 February 1888 he wrote to Loring in what seems to be an attempt at partial justification and an explanation of his activities:

> You may be curious to know why I take such pleasure, such real delight in making these little purchases for you or the Boston Museum. In a general line, when I can show a little gratitude for the glorious time I had in the States, I welcome the occasion. But there is also a personal selfish consideration besides! The few dollars I have spent, on your behalf, have made me learn more about new finds, secret finds, clandestine finds, than I should have learned by myself in five years! I have discovered places and dealers of the existence of which I never had a suspicion; and better than all, I have now the confidence of these men. I am no more an

> official of the State, a public Inquisitor etc.; no. I am a Collector myself! The consequence is that they dont *[sic]* lie any more about the origin and the 'provenance' of the objects and that, as a rule, I am the first to know what happens underwater. You see, I am, *votre obligé* (Lanciani to Loring, 27 February 1888, author's emphasis).[39]

The archaeologist really seems to identify himself as a purchasing agent. Maybe he did so as a *divertissement*. But it must be taken into account that this kind of activity allowed him to deal face to face with dealers and to learn more, not only about the antiquarian market, but also about the trafficking and looting of antiquities, even though it is not easy to reconcile this trading activity with Lanciani's role as a state official of the Ministry of Public Education. Aware of this conflict, he tried to minimise his role in an attempt to defend himself. It is very likely, though, that his scientific work benefited from the kind of information he talks about in his letter.

Lanciani was not only the intermediary for the archaeological objects he acquired, but also – as a well known scholar – he provided important scientific data and information. He put a sort of seal of knowledge on the 'immutable mobiles' which were going to be exhibited in American museums.

In a role play between the Italian scholars – who were creators and mediators of a knowledge related to the archaeological objects found in Italy – and the American museum curators – who had to spread that knowledge by exhibiting those objects – a new idea of 'museum' was created. As Livingstone writes:

> While its architecture was intervening in the cultural struggles of late Victorian society, the museum as an institution did much to promote what has been called an object-based approach to knowing in the decades around 1900, not least in the United States. In a period when Chicago's Field Museum, the American Museum of Natural History, Harvard University's Peabody, and a host of similar institutions came into existence, the idea that knowledge could reside in material objects as much as in texts gripped the imagination of American intellectuals. Appropriately, apologists for museology urged that what distinguished their efforts from those of their antebellum predecessors was precisely that in the new museum specimens were viewed as objects of scientific scrutiny, not simply as spectacle (Livingstone, 2003: 38).

Conclusion

Considering the extraordinary acquisitions that the Metropolitan Museum in New York and the Museum of Fine Arts in Boston went on to make in later years, the bronzes donated by Marquand and the items purchased through Lanciani seem quite unimpressive in comparison (Tomkins, 1989: 87–92).[40] It is clear that the value of the purchases made through the mediation of Helbig and Lanciani is minimal, in terms both of the number of objects purchased and their artistic or archaeological value. Accordingly, where does the importance of those purchases lie?

Those first shipments of antiquities to the United States – which had been made possible as a result of the interest and the contacts of museum directors and curators such as Marquand in New York, Robinson and Loring in Boston and French and Hutchinson in Chicago – made it clear that it was possible and relatively easy to buy authentic works of art and transport them to the United States. As a consequence, the idea that acquiring collections was open not only to students and experts, but also to the general public – whose interest in Antiquity was growing fast – started to develop. Soon, American collectors and curators would no longer be satisfied with small, ordinary objects: they started to look for masterpieces. It was the period defined by Vermeule as 'The Era of the Titans' (Vermeule, 1981: 14): wealthy Americans who owned the greatest companies and banks of the country became passionate about ancient art. In those years many private collections were put together, and many of them were created to ultimately be donated to museums or become museums in their own right. Isabella Stewart Gardner in Boston, John Pierpont Morgan in New York, Edward Waldo Forbes and Paul J. Sachs in Cambridge, Massachusetts, William and Henry Walters in Baltimore, James Deering in Miami, William Randolph Hearst and his mother Phoebe Apperson Hearst in Los Angeles, are only some of the people who spent huge sums of money to buy works of art for their collections which are now open to the public (Johnston, 1999; Strouse, 2000; Chong, Lingner and Zahn, 2003; Orcutt, 2006; Rybczynski and Olin, 2007; Levkoff, 2008; Higonnet, 2009). This new philanthropic activity made it possible for the museums' curators to buy archaeological objects of great value which flooded into the Roman antiquities market after the Banca Romana bankruptcy scandal and the subsequent financial crisis. The era of big acquisitions meant that great American museums such as the New York Metropolitan Museum or the Boston Museum of Fine Arts could compete with their European counterparts.

The American museums began to manage their acquisitions by the mediation of salaried agents in Rome who made purchases on their behalf. This network made the transactions easier, allowed faster connections and facilitated profitable deals. The time of amateur intermediaries had come to an end. The hope of Director-General Fiorelli that objects found in Italy could be exhibited in foreign museums not as 'refugees', but as 'settlers' of the local culture was eventually dead. Helbig and Lanciani acted as intermediaries, following this somewhat questionable criterion of diffusion of knowledge within a network in which scholars were in the middle between collectors and antiquities dealers; but what happened shortly after completely changed the system.

Over the following decades, this new course completely shook up the entire Italian antiquarian market, which was transformed. Huge economic investments, a growing public interest in authentic works of art as well as an increasingly aggressive policy of acquisitions focusing on first-rate items contributed to American collectors dominating the European antiquities scene. The tone of the Italian State officials changed and Aldo Nasi, the Minister of Education, when he was informed about the Rogers bequest to the Metropolitan Museum, wrote to Tommaso Tittoni, the Italian Minister of Foreign Affairs:

> If all this is true, this Ministry must be very concerned. The daily battle – often painful and always difficult – that is being fought in order to wrest what remains of our national artistic heritage out of the hands of the greedy speculators, will now become more arduous and more difficult in the face of huge funds at the disposal of the Museum of New York. To avoid being overwhelmed, the cooperation of all the Italian authorities will be necessary in order to work together to prevent, as far as we can, the outrage that they want to do to our Italy, depriving it of its most precious wealth, for which it holds an undisputed primacy in the world (Nasi to Tittoni, 26 October 1901).[41]

Notes

1 'Come fa meraviglia che vi sia stato un acquirente di detti vasi a meno di supporre che fossero stati destinati per qualche museo Americano, poiché ognuno sa che gli americani, con poca conoscenza delle cose d'arte, e senza muoversi da casa loro, comprano gli oggetti d'arte per commissione; e si fidano della conoscenza in materia archeologica delle persone da loro incaricate di tali acquisti. Se ciò fosse è da deplorare, che persone incaricate di studiare e conservare i monumenti nazionali, o gli oggetti d'arte, ne procurino invece la esportazione all'estero. Come pure non è da lodarsi il far pagare ai forestieri (siano pure americani), oggetti scadenti come oggetti di arte o come monumenti aventi un vero valore o interesse archeologico.' (Archivio Centrale dello Stato, Ministero della Pubblica Istruzione, Direzione Generale Antichità e Belle Arti, I versamento, busta 421, fascicolo 59–48, letter from B. Chigi to G. Fiorelli, 12 December 1889. Translation by the author).

2 'Gli stessi stranieri, i quali ci contendono il possesso degli oggetti nostri, hanno bisogno di quella sanzione legale che la sola autorità governativa può dare, e che serve per poter giudicare meglio del valore scientifico degli oggetti medesimi.' (Fiorelli, 1885. Translation by the author).

3 The Metropolitan Museum of Art (MMA) – Archives – Correspondence; Marquand, Henry Gurdon, 1819–1902. The letters, re-organised by A.M. Del Collo, are available as part of the digital collection of the Thomas J. Watson Library at www.metmuseum.org/art/libraries-and-research-centers/watson-digital-collections. The correspondence includes both Helbig's letters in German and Arthur Frothingham's in English. There are also Francesco Martinetti's letters in Italian; but Marquand's letters are missing. At the time they were received by the Museum, an English translation was provided for some of the letters in German and the quotations that follow in this chapter are taken from those translations.
4 The list of objects donated by Marquand is kept in MMA – Archives – Office of the Secretary Correspondence files 1870–1950 – Marquand, Henry Gurdon – Gift Bronzes 1897 Greek-Roman-Etruscan 1892, 1897–98, N 348, letter from H.G. Marquand to the trustees, 14 May 1897.
5 MMA 97.22.25.
6 MMA 97.22.24.
7 MMA – Archives – Henry Gurdon Marquand Papers, Letter from Wolfgang Helbig to Henry Gurdon Marquand, 22 October 1887, box 1, folder 08, items 1 and 36 (translation).
8 MMA – Archives – Henry Gurdon Marquand Papers, letter from Arthur Frothingham to Henry Gurdon Marquand, 21 June 1889, box 1, folder 8, item 17.
9 MMA – Archives – Henry Gurdon Marquand Papers, Letter from Wolfgang Helbig to Henry Gurdon Marquand, 2 May 1896, box 1, folder 8, item 35.
10 The inventory numbers for these items go from MMA 97.22.1 to MMA 97.22.25.
11 MMA 97.22.11.
12 MMA 97.22.23.
13 MMA 97.22.6.
14 MMA 97.22.4.
15 MMA 97.22.16 and MMA 97.22.17.
16 MMA 97.22.10.
17 MMA 97.22.8.
18 Among the objects sold at the 1903 auction which had been purchased through the mediation of Helbig we find: a middle-Corinthian black-figure *amphora* from La Tolfa near Civitavecchia (Kirby, 1903, n. 972); a red-figure *rhyton* in the shape of a deer's head from Taranto bought by the Metropolitan (MMA 03.3.2; Kirby, 1903, n. 959); a small, marble, helmeted head of Athena resembling the *Athena Giustiniani* (Kirby, 1903, n. 985); a small marble head of Triton (Kirby, 1903, n. 984); a fragment of a marble votive relief with a woman presenting a child and a building in the background (Kirby, 1903, n. 980); a mosaic panel with four masks (Kirby, 1903, n. 1207), and a bronze group with Pan extracting a thorn from a nymph's foot, bought for 3,000 francs and sold for US$2,200 (Kirby, 1903, n. 998).
19 Lanciani's major works include the *Storia degli Scavi di Roma e notizie intorno le collezioni romane di antichità* (Lanciani, 1902–12) and some

popular books published in English such as *Ancient Rome in the light of recent discoveries* (Lanciani, 1888) and *The Destruction of Ancient Rome: A Sketch of the History of the Monuments* (Lanciani, 1899) while his most famous and monumental publication is the *Forma Urbis Romæ* (Lanciani, 1893–1901).

20 The records of the investigation have been partially published (Barnabei and Delpino, 1991: 453–79) and they are kept in the archives of the Biblioteca di Archeologia e Storia dell'Arte (BIASA) in Rome among the Barnabei Papers.

21 The correspondence is kept in the archives of the Boston Museum but only the letters written by Lanciani have been preserved. I welcome the occasion to thank Christine Kondoleon, Senior Curator of Greek and Roman Art of the Boston Museum of Fine Arts (MFA) for giving me the chance to consult the documents.

22 MFA – Department of Art of the Ancient World Archives – letter from Rodolfo Lanciani to Charles G. Loring – 6 January 1888. The Fundilius from Nemi was purchased by Jacobsen and it is now at the Ny Carlsberg (Johansen, 1994, n. 79).

23 MFA – Department of Art of the Ancient World Archives – letter from Rodolfo Lanciani to Charles G. Loring – 30 May 1888.

24 MFA – Department of Art of the Ancient World Archives – letter from Rodolfo Lanciani to Charles G. Loring – 16 February 1888.

25 In all, 187 objects were purchased through the mediation of Lanciani (10 of which were later de-accessioned), plus 168 coins (64 later de-accessioned). The online collection database associates Lanciani's name also with the sale of a bronze bust of Augustus (MFA 90.163) which Samuel D. Warren (Edward Warren's brother) donated to the museum in 1890 and of an *amphora* (MFA 41.913) donated by Edward Jackson Holmes in 1941.

26 Terracotta statuettes (MFA 88.353–88.364) purchased from Pennelli at a cost of 600 French francs; see Borsari (1886). On the materials from Caere in the MFA, see Nagy (2008).

27 Augusto Alberici was an art dealer and his shop was situated in via dell'Olmata 52 in Rome. MFA – Department of Art of the Ancient World Archives – letter from Rodolfo Lanciani to Charles G. Loring – 27 February 1888.

28 A bust of Tiberius from Civita Lavinia (today identified as Tiberius' brother Drusus) (MFA 88.346), a bust of Caligula from Porta Salaria bought from the art dealer Eliseo Borghi and today believed to be a portrait of a young man from the Trajan era (MFA 88.348) and a bust reworked as Emperor Balbinus bought from Eliseo Borghi and said to have been found in the quarter of Villa Ludovisi (MFA 88.347); a Julio-Claudian bust (now identified as Herakles MFA 88.350) purchased from Augusto Valenzi for 200 French francs; an Artemis Colonna (MFA 88.351) bought from Alessandro Fausti for 145 French francs and said to have been found in 'località Vigna del Torrione' in Grottaferrata and one bust of a woman identified with Diana (MFA 88.352), purchased from Scalambrini.

29 The 168 coins dating from Julius Caesar to Justinian cost 262.75 French francs.

30 These purchases consisted of pottery from a tomb on the slope of Monte Cucco bought from art dealers Fausti and Alberici (MFA 88.538–88.550); pottery from a grave in the Esquiline region bought from Jandolo (MFA 88.551–88.555 and 88.610–88.612); terracottas and bronzes from the Sanctuary of Diana Nemorensis discovered by Luigi Boccanera in 1887 (MFA 88.556–88.562 and 88.613–88.628): see Robinson (1889); terracottas from Caere bought from Pennelli (MFA 88.613–88.628): see n. 35; specimens of terracotta lamps (MFA 88.571–88.582); samples of Arezzo ware, mostly from the Gardens of Caesar (MFA 88.583–88.608); Roman architectural elements, so-called 'lastre Campana' (MFA 88.563–88.570); Roman inscriptions and bas reliefs (MFA 88.631–88.637).

31 This series consisted of a head of Caius Memmius Caecilianus from Piazza dell'Esquilino bought from Jandolo for 200 *lire* (MFA 88.349: see Gatti, 1887, in part p. 179); a portrait of a Roman in the Republican veristic style (MFA 88.638, de-accessioned); a bust of Domitian from Tusculum (MFA 88.639); a portrait of Julia as Artemis (MFA 88.641, de-accessioned); a Julio-Claudian portrait of a girl (MFA 88.642); a little Greek Venus (MFA 88.640, de-accessioned); a head of an African man (MFA 88.643); a bust of Ajax (MFA 88.644, de-accessioned).

32 MFA 89.9–89.31.

33 The series consisted of a head of Mercury with the *petasos*, now considered a forgery (MFA 89.2); a head of a faun (MFA 89.3); a bust of the Emperor Maximin (MFA 89.4); a head and a bust of Domitian (MFA 89.5 and 89.6); a bust of Minerva (MFA 89.7) and an idealised female head, perhaps a Muse, according to Lanciani (MFA 89.8).

34 They were vases (MFA 89.256–89.263; 89.266a–b; 89.268; 89.269a–b; 89.272; 89.273; 89.275; 89.561 and 89.562) purchased from Pio Marinangeli and allegedly found at Corneto Tarquinia, Capua, Nola and Ruvo.

35 Including a red-figure *crater* from Nola (Art Institute of Chicago, 1889, n. 301; ARTIC 1889.16) and a black-figure vase from Orvieto (Art Institute of Chicago, 1889, n. 341; ARTIC 1889.10)

36 Anon (1889, nn. 364, 366, 367, 368, 369, 370, 371, 375, 376, 377, 378, 379). The Art Institute of Chicago. Palombi also identifies (Anon, 1889, n. 363) a terracotta statuette of a woman (Palombi, 2006: 134–5, n. 192).

37 Palombi identified, among the Lanciani Papers at the BIASA, a letter from Cesnola to French which the latter forwarded to Lanciani. Cesnola was seeking advice on the purchases and French sent a reply on 22 March 1890: 'The few marble sculptures from Rome which we possess were bought from dealers in Rome: I suppose the New York Museum would have no difficulty in picking up a few good objects in the same way', and further, 'I carried a letter of instructions to Sig. Lanciani from his sister-in-law, who lives here, which led him very kindly to give some advice about the purchases. I felt grateful, and expressed it in the catalogue. The greater part of our collection came from Naples, where it was inspected by another expert. I do not know Mr Helbig, but I should expect that either he or Sig. Lanciani would be willing to assist any American Museum in the same way the latter

did us' (Palombi, 2006: 138–9, footnote 196). Furthermore some letters written by Lanciani to Hutchinson are preserved at the Newberry Library in Chicago dating from April 1889 to August 1892 (Newberry Library Archives – Charles L. Hutchinson Trustees President Correspondence – F–Z 1883–1924, Box 1).

38 A letter of defence written by Martin Brimmer, President of the Trustees of the MFA, was attached to the statement (Palombi, 2006: 137, n. 195).

39 MFA – Department of Art of the Ancient World Archives – letter from Rodolfo Lanciani to Charles G. Loring – 27 February 1888.

40 In 1901, the Metropolitan Museum received a donation of US$5 million from the railroad magnate Jacob S. Rogers. This allowed the purchase of artworks of incredible value, such as the Roman frescoes from Boscoreale, the Monteleone chariot and the Giustiniani marbles. In 1903, thanks to the intermediation of Warren, a seated female statue without its head (MFA 03.749) from Vasciano (Todi) arrived in Boston (Pasqui, 1900) while in 1909 the 'Boston Throne' (MFA 08.205) finally reached the Museum of Fine Arts.

41 'Se tutto ciò è vero, questo Ministero dev'essere ben preoccupato. La lotta quotidiana, spesso penosa, difficile sempre, che si combatte per strappare dalle mani degli avidi speculatori quel che ancora rimane del nostro patrimonio artistico nazionale, diventerà ora, di fronte alle ingenti somme di cui può disporre il Museo di New York, più aspra e più difficile; e per non restare sopraffatti, occorrerà la cooperazione di tutte le autorità italiane, aspiranti ad un solo intento, quello di prevenire, fin quando è possibile, l'oltraggio che si vuol fare all'Italia nostra, spogliandola della ricchezza più cara e per la quale tiene nel mondo un primato indiscutibile.' (Archivio Centrale dello Stato, Ministero della Pubblica Istruzione, Direzione Generale Antichità e Belle Arti, III versamento II parte, busta 323, fascicolo 613–8, letter from A. Nasi to T. Tittoni, 26 October 1901. Translation by the author.)

4

Digging dilettanti: the first Dutch excavation in Italy, 1952–58

Arthur Weststeijn and Laurien de Gelder

What determines the possibility of an archaeological excavation abroad and its success? In September 1952, when two Dutch archaeologists with little experience on the ground started digging underneath the Santa Prisca church on the Aventine hill in Rome, this seemingly trivial question loomed large over their pioneering efforts. For decades, Rome had been the obvious centre of all archaeological attention worldwide – but the Eternal City was essentially off limits for non-Italian campaigns. Ever since the unification of Italy, with Rome becoming the new nation's capital in 1871, large-scale fieldwork projects in the city had been effectively restricted to Italian experts, the likes of Rodolfo Lanciani, Giacomo Boni and Alfonso Bartoli. But after the collapse of the Fascist regime and with the nascent process of European collaboration in the aftermath of the Second World War, Italy gradually opened up its rich soil to archaeologists from abroad. The Dutch were among the first to profit from this opportunity; their fieldwork project at Santa Prisca, though minor in scale, can be seen as the first foreign campaign in Rome since the famous French excavations on the Palatine hill during the reign of Napoleon III in the 1860s. How did this project emerge, and what circumstances made it possible?

In this chapter we take the largely unknown but highly significant example of the Dutch excavation at Santa Prisca to offer an historical contextualisation of the networks (personal, professional and political) that impact upon archaeological practice in an international setting.[1] We thereby aim to show how archaeological knowledge is produced through the interaction of individual and collective processes of networking that develop within specific geographies of knowledge (in our case the particular institutional setting of foreign institutes in Rome) and through both structured collaborations and informal conversations between archaeologists and non-archaeological actors. The data we present especially reveal to what extent scholarly concerns about archaeological inexperience can be overridden by political interventions based upon such institutional–informal networking.

The two protagonists of our story are the Dutch archaeologist Carel Claudius van Essen (1899–1963) and his colleague and compatriot Maarten Vermaseren (1918–85). For about a decade, from 1952 to 1958 and again from 1964 to 1966, they carried out fieldwork underneath and in the area surrounding the Santa Prisca under the aegis of the Royal Netherlands Institute in Rome (KNIR) (see Figure 4.1). While other nations competed for access to large-scale excavation sites in Italy such as Greek colonies, Etruscan settlements and Roman provincial towns, the Dutch started digging under a humble *titulus* located in the heart of the Eternal City. On the Aventine hill, they succeeded in establishing a fruitful and harmonious collaboration with the Italian archaeological services, which resulted in important findings, especially in the famous third to fourth century *mithraeum* and parts of a Roman habitation from the imperial period underneath the church. The results of the first phase of the excavations were extensively published in 1965 (Vermaseren and Van Essen, 1965), while the findings of the second phase of the fieldwork (1964–6) that concentrated on the garden of the church have recently been investigated afresh in an analysis of the excavation's legacy data (Armellin and Taviani, 2017; Kruijer, Hilbrants, Pelgrom and Taviani, 2018). These data include Vermaseren's personal archive, containing around forty notebooks with field reports, architectonic and stratigraphic descriptions, sketches and more than a thousand photos, as well as the complementary archive of Rome's archaeological service, which contains a rich collection of institutional correspondence, maps and decrees related to the Santa Prisca excavations.[2] These archives not only allow for a thorough reconstruction of the two phases of the excavation, but also hold the key to the understanding of the dynamics of the social, cultural and political contexts within which Dutch archaeological fieldwork abroad took wing in the 1950s.

Our approach to these diverse legacy data starts from the perspective that they contain information about the historical development of archaeological practice, thus making the non-archaeological context of the excavation itself into a subject of investigation. Recent work on the evolution of archaeology into an independent scientific discipline has covered many approaches to studying the histories of archaeological knowledge production, resulting in biographies of discoverers, genealogies of discoveries and historical analyses of the institutional contexts

4.1 Maarten Vermaseren (left) and Carel Claudius van Essen studying the portrait of Serapis, found in the *mithraeum*. Vermaseren archive, Royal Netherlands Institute in Rome.

in which archaeological knowledge is produced (Murray and Evans, eds, 2008; Murray, 2012). Apart from the many hagiographies of great archaeologists, particular attention has been paid recently to the political dimensions of archaeology and the manifold ways in which archaeology has been used for political, especially nationalist, purposes (David, 2002; Dyson, 2006; de Haan, Eickhoff and Schwegman, eds, 2008). In the context of Italian archaeology, a clear example of this trend is the increasing focus on the use of archaeology during the Fascist regime as a political tool and as an instrument of *romanità* (Arthurs, 2012, 2015).

Much less is known, however, about the post-war period, in which the nationalist perspective on the past gradually gave way to a more internationalist practice of archaeological fieldwork on Italian territory. In this chapter, we focus on this crucial period, zooming in on the Dutch excavation at Santa Prisca to tell a larger story of the disciplinary infrastructure and the social, cultural and political contexts in which archaeological practice developed in Italy after the fall of Fascism. The Santa Prisca case is especially interesting because it shows how two archaeologists from a small country with little personal field experience managed to undertake and accomplish a difficult excavation campaign in the centre of Rome, the heartland of antiquity. We apply actor-network-theory (Latour, 1996, 2005) to argue that their relative success in doing so can be explained through the various formal, informal and institutional networks within which they operated, which resulted in attracting significant private and public funding as well as effective archaeological valorisation. Specific attention is paid to the combination of individual and institutional networking, which reveals the agency of single archaeologists and non-archaeological actors (such as politicians) as well as that of institutions. In this context, we use insights based on the geography of knowledge (Naylor, 2002, 2005) to show how the local environment of foreign institutes in Rome provided a favourable institutional framework that determined the possibility and success of this particular archaeological excavation outside the home country.[3]

Before the excavation: institutes and individuals in the opening up of Italian archaeology

From the last quarter of the nineteenth century to the end of the Second World War, Italian Classical archaeology went hand in hand with a clear nationalist agenda. The young Italian state, crowned with the conquest of Rome in 1870, established new ministerial bodies that organised a hierarchical regulatory framework for archaeological practice and preservation throughout the peninsula. Under the directives of the central Direzione Generale delle Antichità e Belle Arti (Directorate-General for Antiquities and Fine Arts), a subsidiary of the Ministry of Education, large-scale excavation campaigns were organised in Rome and elsewhere, progressively excluding foreign archaeologists in the practice of archaeological fieldwork. This *de facto* boycott of non-Italian archaeologists remained in force for decades and was further intensified during the Fascist regime, when archaeology became one of the main vehicles of Mussolini's pretended rebirth of the Roman Empire (Bourdin and Nicoud, 2013; *cf.* Palombi, 2006).

With the fall of the regime in 1943 and the subsequent liberation of Italy in 1945, this attitude started to change. In the years following the war, many members of the Italian intellectual and academic community turned their back on the ideological and practical remnants of Fascism, and Italian Classical archaeologists (many of whom had served under Fascism) looked for new approaches to avoid the mixing of politics and archaeology. As a result, much emphasis was placed on technical studies that carried no ideological connotations (Barbanera, 1998). From an international perspective, the overall post-war spirit was one of international collaboration, illustrated by the opening moves of European collaboration and the *rapprochement* with the United States of America that culminated in the Marshall Plan. In the archaeological community in Rome, this spirit of international collaboration had an especially profound impact upon the institutional framework of the foreign schools, the various academies and institutes of non-Italian academic communities that had been established in the late-nineteenth and early-twentieth centuries in the context of the nationalist competition over Rome and antiquity. Immediately after the end of the Second World War, two important organisations were created that instead focused deliberately on collaboration and shared international interests. In 1945, the Associazione Internazionale di Archaeologia Classica (AIAC) was founded, with a leading role for Erik Sjöqvist, the director of the Swedish Institute in Rome. The AIAC expressly aimed to exchange and disseminate archaeological discoveries in Italy, and it was entrusted with the restitution to Rome of the four German libraries which had been shipped back to Germany during the war. A year later, in the same collaborative spirit as the AIAC, the Unione degli Istituti di Archeologia, Storia e Storia dell'Arte (Unione) was established, which likewise meant to strengthen the relationship between the foreign schools (Whitling, 2018; *cf.* Rietbergen, 2012).

In this climate of scientific exchange and international collaboration, the Italian government gradually became more hospitable to foreign archaeology on Italian territory. The establishment of organisations such as the AIAC and the Unione strengthened the perception of the cultural relevance of Classical archaeology amongst Italians and other nationals alike. Meanwhile, after decades of exclusion of non-Italian archaeologists, the foreign schools were eager to take advantage of the sudden increase of archaeological opportunities; despite the spirit of international collaboration, a large-scale fieldwork project was still essentially seen as a matter of national academic prestige. In the immediate post-war years most of the foreign schools (and their respective national governments) were ill prepared to fund archaeological fieldwork, but soon the big players took advantage of Italy's renewed hospitality, with the French starting an excavation at Bolsena in 1946 and the Americans in 1948 at the Roman colony of Cosa. The next year, Belgium also received permission to excavate the site of Alba Fucens. Even small countries now joined the hunt for prestigious archaeological fieldwork projects in Italy (Linde, ed., 2012b; Bourdin and Nicoud, 2013).

The KNIR followed this general trend. The Institute had been established in 1904, mainly as an institutional home for Dutch historians coming to Rome to study the Vatican archives. Before long, the Institute also embraced other scholars from different disciplines, and in 1920 Hendrik Leopold (1877–1950) was appointed as an archaeological assistant. Trained as a journalist, Leopold reported archaeological developments and discoveries to the academic community affiliated to the Institute and the general Dutch public via various newspapers (Cools and De Valk, 2004: 55–6).[4] But Leopold was a trained archaeologist himself, having been a member of the pioneering excavation team of the Dutch archaeologist Carl Wilhelm Vollgraff (1876–1967) in Argos, Greece at the beginning of the twentieth century. During Easter 1932, at the invitation of the palaeontologist Ugo Rellini Leopold was even allowed to join the Italian excavation in Northern Apulia for a few days (Heres, 1989: 82). A Dutch-led dig in Italy, however, was not possible until after the Second World War.

With the retirement of Leopold in 1942, the long-standing director of the KNIR, Godfried Hoogewerff (1884–1963), started looking for a new archaeologist who could also fill the role of vice-director. At the recommendation of Gerard van Hoorn (1881–1969), who taught archaeology at the University of Utrecht, Hoogewerff appointed Carel Claudius van Essen. After his graduation in Classical Languages in 1921, Van Essen had participated in excavation projects in The Netherlands and he worked as an assistant curator in the archaeological museum of Constant Willem Lunsingh Scheurleer (later appointed as professor in Greek Archaeology at Leiden University).[5] The Scheurleer family went bankrupt in the crisis of the 1930s and the museum had to close, but this experience made Van Essen an important member of the burgeoning archaeological intellectual community in The Netherlands.[6] Moreover, Van Essen was very familiar with Rome, having been granted scholarships from the Netherlands Institute several times during his studies. These research visits allowed him to start building a professional network in Rome, including the well-known Italian Etruscologist Antonio Minto and his student Ranuccio Bianchi Bandinelli. Encouraged by

Hendrik Bolkenstein, a specialist in ancient religion at the University of Utrecht, Van Essen finished his dissertation on Etruscan tomb painting in 1927 (Cools and De Valk, 2004: 93–5).

In 1947 Van Essen returned to Rome for good as archaeologist and vice-director at the KNIR, having survived the war years fighting in the anti-German resistance. Soon after his arrival in Rome, he was determined to build new personal and institutional ties that could strengthen the position of the Institute through informal as well as institutional platforms for knowledge creation. As early as July 1947, for two weeks Van Essen joined the excavation at the Etruscan site of Bolsena with the École française *de* Rome (Heres, 1989: 82). A year later, he joined the board of the AIAC, together with the retired Leopold; he consequently attended its meetings and also became a member of the editorial board of *Fasti Archeologici, Annual Bulletin of Classical Archaeology*, the association's journal. During this period, his interests shifted from Etruscology to topographical and architectonic studies related to early Christian and Byzantine history and archaeology (Van Essen, 1950).

The post-war spirit of international collaboration (for the sake of national prestige) was also shared by Hoogewerff, who as director served as the Dutch representative on the board of the Unione. After the war, Hoogewerff seized the opportunity presented by changing circumstances to increase the cultural and professional reputation of the KNIR. To this end, he standardised the allocation of scholarships for researchers, and one of the first people to be granted a scholarship, in 1946, was Maarten Vermaseren. Like Van Essen trained as a Classicist, Vermaseren was a disciple of the famous Belgian scholar Franz Cumont (1868–1947), under whose direction he studied the sanctuaries of the Roman Mithras cult – *mithraea* – in Rome and Ostia. In 1947, while Vermaseren was still in Rome, Cumont died, and his pupil was regarded as the rightful person to succeed Cumont in continuing the studies of ancient 'oriental' religion in the Mediterranean (Roos, 1950–51).[7]

With the more or less simultaneous appointment of Van Essen and rise of Vermaseren in 1947, the Netherlands Institute in Rome suddenly increased its archaeological standing and potential. Meanwhile, diplomatic ties between Italy and The Netherlands were strengthened in a general climate of west European reconciliation. The young Catholic historian Jan Poelhekke (1913–85), the Institute's new director from 1950 onwards, played a key role in this process. In 1951, Poelhekke was appointed cultural attaché to the Dutch Embassy in Rome with the task of serving Dutch cultural interests in Italy, and in the same year a cultural treaty was signed between The Netherlands and Italy. Poelhekke, in his double role as director of the KNIR and cultural attaché to the Embassy, became the spider in the web of this nascent

Italo-Dutch cultural collaboration (Cools and De Valk, 2004: 88–96). The conditions for starting a Dutch excavation in Italy, academic as well as political, had never been this favourable.

Starting an excavation abroad: academic diplomacy between Rome and The Hague

On 6 March 1947, Vermaseren guided a small Dutch delegation through the rooms of the *mithraeum* underneath the Santa Prisca church on the Aventine hill (*Mededelingen van het Nederlands Instituut te Rome* (MNIR) [26] 1950: xx).The cult site for the deity Mithras had been partly excavated in the 1930s by the Augustinian friars of the church;

4.2 The cult niche of the *mithraeum* of the Santa Prisca with Mithras killing a bull (*tauroctony*). Vermaseren archive, Allard Pierson Museum (Amsterdam).

4.3 In 1948 Maarten Vermaseren commissioned the Italian architect L. Cartocci to sketch the *mithraeum* of the Santa Prisca. Vermaseren archive, Allard Pierson Museum (Amsterdam). Copyright © Allard Pierson Museum (Amsterdam).

their efforts had brought to light a cult-niche with a unique stucco decoration of the sun god Mithras as well as remarkable frescoes and inscriptions related to the cult in the *spelaeum*, the central room of the *mithraeum* (see Figure 4.2) (Ferrua, 1940a, 1940b). During the tour and in a subsequent report published in the MNIR, the publication series of the KNIR, Vermaseren could not resist expressing his concerns about the state of conservation of the frescoes in the damp rooms under the Santa Prisca. He argued: '[b]ecause of the deterioration of the frescoes, it would be desirable to photograph them in colour print, but it is necessary to clear the surrounding areas from their filling' (MNIR [26] 1950: lxxii). This statement is the first indication of Vermaseren's interest not only in safeguarding the frescoes, but also in further exploring the surrounding areas of the *mithraeum* (see Figure 4.3).

The Italian authorities shared that interest. In September 1943, a few years before Vermaseren's tour of the Santa Prisca, Bianca Maria Felleti-Maj, an assistant at the Italian archaeological service, had also noticed the deterioration of the frescoes. Her inspection of the *mithraeum* resulted in a list of suggestions to preserve and restore the archaeological site, emphasising the importance of conservation of the frescoes, which were degenerating because of humidity and the bad state of the rooms of the *mithraeum* (ASSAR, busta 40.6). Felleti-Maj wrote her report days after the German occupation of Rome, and because of the war there was no practical follow-up to her alerting memo. But eventually the issue was brought up again in October 1951 in an internal report of the Italian archaeological service by Carlo Cecchelli, professor of Christian archaeology at the University of Rome (and one of the many local academics who had vividly supported Fascism in their earlier career). In his report, Cecchelli stated that the *mithraeum* and the adjacent structures of an imperial habitation should be further examined, for which a specialist was needed with knowledge and experience in Classical and late Roman archaeology. One of the expressed aims of the conservation of the *mithraeum* was that the archaeological site could be exploited for tourism (ASSAR, busta 417.8, 1c).

Now was the moment for Vermaseren and Van Essen to act. Their cooperation in the matter came naturally, as they shared a similar academic background in Classical languages and the same interests in late ancient history and archaeology, especially in the field of late Roman religion. From a practical point of view, Vermaseren's professed ambition to start an excavation underneath the Santa Prisca church was much easier to realise under the aegis of an institutional infrastructure in the person of Van Essen as vice-director of the KNIR. Accordingly, the two archaeologists joined forces and decided to take up Cecchelli's suggestion that the site be excavated and restored, which had probably been advertised in the circles of the AIAC and the Unione. In March 1952, some five years after Vermaseren's first tour of the Santa Prisca and almost a year after he obtained his PhD, the Direzione Generale delle Antichità e Belle Arti received an official letter from Van Essen, who proffered himself and Vermaseren as the best candidates for safeguarding (by detachment) the frescoes of the *mithraeum* and for excavating the surrounding areas of the cult site (ASSAR, busta 417.8, 1c).

The Dutch proposal entered the maelstrom of bureaucratic negotiations between different Italian ministries and the archaeological service, which had to decide whether Van Essen and Vermaseren were suited for the job. After a few weeks, Salvatore Aurigemma, the head of the archaeological service, gave his authoritative verdict on the matter, writing to the Ministry of Education that he 'recommends welcoming the proposal very favourably'. However, Aurigemma could not hide his concern about the expertise of the Dutch team, adding with academic understatement that he 'was not aware of the specific technical knowledge on the matter of Vermaseren, to whom the Netherlands Institute intends to give the direction and the responsibility for the enterprise' (ASSAR, busta 275.4). Apparently, this concern was not seen as decisive, and in June 1952 Van Essen and Vermaseren received official permission to start the first Dutch excavation on Italian soil, under the condition that the Istituto Centrale del Restauro and the Italian archaeological service would supervise the restoration and the fieldwork (ASSAR, busta 275.4).

The swiftly made decision to entrust the excavation under the Santa Prisca to Vermaseren and Van Essen betrays the extent to which the post-war climate of collaboration and cultural exchange, recently decreed in a cultural treaty between Italy and The Netherlands, prevailed over concerns relating to scholarly expertise and experience. Indeed, Van Essen and Vermaseren had been trained as Classicists, and although they had gained some know-how in earlier fieldwork projects, they did not have any experience in directing a challenging excavation in the midst of the stratigraphic complexity of Rome. Clearly, the overriding factor was the academic prestige of Van Essen and Vermaseren, both respected scholars with a broad international network in the archaeological community in Rome. This prestige was strengthened further by the institutional backing of the Netherlands Institute and its director Poelhekke. Also on a higher diplomatic level, the excavation was born under a lucky star: the new Dutch ambassador in Italy appointed that year, Han Boon (1911–91), was an historian by training, with a PhD from Leiden University (supervised by Johan Huizinga) and a genuine interest in archaeology. Together with Alexander Byvanck, professor of Classical Archaeology at Leiden, Boon founded a small committee that supported the Dutch fieldwork at Santa Prisca among Dutch governmental and academic circles in order to stimulate financial aid (*cf.* Boon, 1989).

The Dutch initiative to start digging in Rome, then, was soon endorsed institutionally and politically. But before the project could take off, another difficulty had to be solved: funding. In recent years, shortage of funding had been one of the major reasons why the Dutch had not yet been able to initiate a costly archaeological campaign in Italy (*Nieuwe Leidsche Courant*, 13 November 1954: 13). Excavating urban archaeological sites such as the *mithraeum* beneath the Santa Prisca obviously costs time and money, as well as demanding a large workforce with expertise in complex stratigraphy and diverse material. Substantial sums were required for personnel and for the complex techniques and tools needed to detach the frescoes and remove the filling of the surrounding areas of the *mithraeum*. But again, private prestige and networks could be mobilised to cover these costs. Given the swiftness of the Italian authorities in permitting the excavation, the Dutch had to operate fast in creating a financial infrastructure to embark upon the project. In order to get extensive funding in a relatively short time, Van Essen and the committee of Boon and Byvanck approached private capital. Buoyed up by their broad social network, they found J.M. Hondius and Ada Hondius-Crone, a wealthy couple from Amsterdam interested in oriental and ancient religions, willing to donate a large sum that allowed a first campaign to begin.

The preparations for that campaign took off in September 1952 and, supported by the promising results, Van Essen and Vermaseren successfully applied for subsequent funding from the Dutch government. Arguing that the Santa Prisca *mithraeum* was 'a wholly unique sanctuary and certainly the most important of all Mithras sanctuaries in the world', they convinced the board of the Netherlands Organisation for Pure Scientific Research (ZWO), which had been established two years before as the national research council that subsidised Dutch research (Archive NWO/ZWO: inv. nr. 31). During this formative period, one of the board members of the organisation was Hendrik van Wagenvoort (1886–1974), a Latinist with a special interest in archaeology and in particular in ancient Roman religion. Van Wagenvoort was a personal friend of Vermaseren and his formal supportive role was crucial for the continuation of the Santa Prisca campaign: from 1953 onwards, Van Essen and Vermaseren could count on substantial and systematic funding provided by ZWO. Moreover, in the heart of political decision-making in The Hague, the team received even more prominent backing from Jo Cals, who was Minister of Education, Culture and Sciences for the Catholic People's Party from 1952 to 1963. Cals was a close friend of the Catholic priest and historian Reinier Post (1894–1968), who in the 1930s had been a member of the scientific staff of the KNIR. On the fiftieth anniversary of the Institute in 1954, Cals attended the celebrations in Rome, honouring the requests of Poelhekke and Post to give further ministerial support to the Institute. From the Roman side, those present at the celebrations included numerous church officials as well as the Italian president Luigi Einaudi (Cools and De Valk, 2004: 80, 92, 99). The Santa Prisca excavation greatly profited from this intensified academic diplomacy (with strong Christian Democratic overtones) between Rome and The Hague. Pivoting on such characters as Van Essen, Boon, Poelhekke, Byvanck, van Wagenvoort, Post and Cals, this was an old boys network if ever there was one.

Valorising an excavation abroad: preservation, founding a museum and media campaign

Once the excavations underneath the Santa Prisca had started and the funding for the project was secured, the real challenge began. For Van Essen and Vermaseren, dealing with a highly complex archaeological stratigraphy in an urban setting was something completely new. By trial and error, they managed to excavate the rooms adjacent to the

4.4 A harmonious cooperation. From left to right: the Italian front man Moreschini, Van Essen, Vermaseren, O. Testa (assistant at Soprintendenza Roma I) and on the right the son of Moreschini (picture by W. van den Enden). Vermaseren archive, Allard Pierson Museum (Amsterdam). Copyright © Allard Pierson Museum (Amsterdam). All rights reserved and permission to use the figure must be obtained from the copyright holder.

mithraeum, assisted by local workmen (and the occasional Dutch student) who performed the hard work. In the process the workforce unearthed many older structures relating to a late-first century habitation on the site, which had been substantially altered in the third century. Given this stratigraphic complexity, Van Essen and Vermaseren succeeded remarkably well in their efforts to make sense of the archaeological data (Vermaseren and Van Essen, 1965).[8] An important factor in that success was the smooth collaboration with Italian specialists in the archaeological service, who had much more experience on the ground (see Figure 4.4). Whilst Vermaseren progressively took the lead in coordinating the campaign, Van Essen employed his personal network and prestige to create a favourable working environment. Significantly, his networking also involved the church authorities: in 1953, Van Essen was present at the ceremony in the Santa Prisca when the titular church was assigned to Cardinal-Priest Angelo Roncalli (MNIR [29] 1957: 10). On that occasion, Roncalli explicitly expressed his support for the Dutch excavations under his *titulus*. His leading role in the church hierarchy was sealed with his election in 1958 as Pope John XXIII.

In this favourable setting, the fieldwork at Santa Prisca advanced. The initial aim of the excavations was the safeguarding of the frescoes, for which the adjacent rooms of the *mithraeum* had to be cleared. Accordingly, the conservation and preservation of the site became the main concern in the correspondence between the Dutch scholars and the Italian authorities. As early as the first investigative campaign in 1952–3, it became clear that detaching the frescoes was a very technical and difficult task. An internal memo of the Direzione Generale delle Antichità e Belle Arti to Pietro Romanelli, the new head of the archaeological service, stated that the humidity should be controlled in the spaces of the *mithraeum*, or else mechanical tools would have to be used under the supervision of a specialist to detach the frescoes. The memo was forwarded to Van Essen, but the advice proved easier given than acted-on (see Figure 4.5) (ASSAR, busta 417.8, 1c). At the end of May 1953, one of the arches collapsed in the rooms that Van Essen and Vermaseren had excavated. Instead of detaching the frescoes from the walls, the first priority now became preserving the existing structures *in situ* so far as possible. Vermaseren told Romanelli that the various possibilities for cleaning the frescoes would be explored as well (ASSAR, busta 417.8, 1c). He started to pay particular attention to the conservation and restoration of the excavated areas and of the *mithraeum* itself, and the field reports, photos and drawings in the Vermaseren archive reveal that his main concern increasingly lay with the architectonic structures of the excavation. The team's objectives were supported by the new influx of Dutch funding from ZWO in the spring of 1953, which permitted fieldwork to continue in the subsequent months and again in the summers of 1954, 1955 and 1956, concluding with a final campaign in the autumn of 1957.

Throughout the excavations, heritage preservation and valorisation remained the priority for the Dutch team and the Italian archaeological service alike. As early as 1953, Vermaseren and Romanelli discussed the possibility of establishing a small *antiquarium*, a modest museum of findings, in the excavated *nymphaeum* of the site (see Figure 4.6). Proposing such a long-term collaborative effort at on-site valorisation exemplified the spirit of Italian-Dutch cultural cooperation that had made

4.5 Maarten Vermaseren studying the state and subject matter of the frescoes in the *mithraeum*. Vermaseren archive, Allard Pierson Museum (Amsterdam).

the excavation possible in the first place. Moreover, the small museum of Santa Prisca also revealed the successful private-public partnership that had accompanied the campaign from the start. The eventual establishment and furnishing of the museum was made possible by several donors in The Netherlands and Italy, including private companies such as Philips (which unsurprisingly funded the lighting for the museum) and, curiously, the Milanese branch of Berkel, a Dutch manufacturer of meat-slicing machines. Once again, van Wagenvoort and the committee of Boon and Byvanck played a leading role in the fund-raising, appealing to the generosity of 'all those who grant our Fatherland an honourable place in the world on a cultural level' (Vermaseren archive: note dated March 1957). At the entrance of the site beneath the Santa Prisca, the financial support of all donors is still commemorated on a marble plaque that celebrates, with a faint hint of such national pomposity, 'the first Dutch excavations in Roman soil'.

The museum was officially inaugurated on 21 May 1958, signalling the Italian-Dutch collaboration over the years and the shared desire to make the site available for tourism and valorisation. In the presence of

4.6 Glimpse of the *antiquarium* presenting the most important finds of the Santa Prisca excavations. Photo collection Anton von Munster, Royal Netherlands Institute in Rome.

a Dutch delegation (which included the Hondius couple as well as van Wagenvoort and other members of the ZWO), the museum and the site were formally handed back to the Italian authorities by ambassador Boon (see Figure 4.7). The year before, the Treaty of Rome had given birth to far-reaching economic integration in western Europe, and the ceremony at the Santa Prisca testified to the rising tide of international cooperation and exchange on the continent. The first Dutch excavation in Italy formally came to an end, and the direct responsibility for the maintenance of the site and the *antiquarium* was returned to the rightful owners of the property, the Augustinian friars of the church, under the supervision of the Italian archaeological service.

4.7 The wife of the Dutch ambassador in Italy, Han Boon, inaugurates the *antiquarium*, revealing the marble slab with the various donors to the Santa Prisca excavations. The marble slab can still be seen in the *nymphaeum* of the *mithraeum*. Vermaseren archive, Allard Pierson Museum (Amsterdam). Copyright © Allard Pierson Museum (Amsterdam). All rights reserved and permission to use the figure must be obtained from the copyright holder.

But with the conclusion of the archaeological campaign, the campaign for publicity was not yet over. From the very start of the excavations, the network of Van Essen at the KNIR had been mobilised to gain attention for the *mithraeum* in the Dutch and international media. In October 1952, Adriaan Luijdjens, the leading Dutch correspondent in Rome who had worked for many years at the Institute as Hoogewerff's private secretary, published an article on the excavations on the front page of the *Algemeen Handelsblad*, one of the most important newspapers in The

Netherlands, heralding the significance of this first Dutch excavation in Italy ([Luijdjens], 1952; *cf.* Haitsma Mulier, 1991). More articles by his hand followed, and in 1958, on the occasion of the inauguration of the museum, Luijdjens wrote a final piece in which he lauded the Dutch campaign not only for its scientific results, but especially for the way the excavation was valorised for future visitors. Luijdjens made it adamantly clear that an archaeological excavation in Rome was not only of academic interest: what was truly at stake was the Dutch national interest, 'waving the cultural flag at the highest level' ([Luijdjens], 1958).

That message was not lost on Vermaseren, who took the lead in the gradual national and international media campaign for the Santa Prisca excavation. Upon leaving for Rome in 1952, he had already written a leaflet on the excavations to inform a wider audience in The Netherlands about the plans for subsequent years. Noticing how 'students of different nations bask in Rome at the cultural hearth of Western European civilisation', he eulogised the KNIR as a 'spiritual ambassador between the fatherland and Roman civilisation'. These remarks betrayed how the post-war climate of international collaboration set the scene for competition between nations – with Rome being both the main arena and the ultimate prize. For Vermaseren, the excavations at Santa Prisca meant that the Dutch had finally joined the game, and that 'our nation, following other countries, could now help to uncover part of the veil of Roman civilisation' (Vermaseren, 1953: 1–2). Rome, the Eternal City, was seen as the matchless marker of universality, a *città madre* for the West; a claim on the city on behalf of the 'fatherland' proved that one's country was a worthy member of Western civilisation and a reliable partner in European unification.

To make clear that the Dutch had become serious players in Roman archaeology, frequent tours were organised during the excavations for colleagues from the other international institutes in Rome. Moreover, an unexpected opportunity to boost international attention for the Santa Prisca site arose with the discovery in 1954 of a comparable *mithraeum* in the centre of London at Walbrook. In *The Illustrated London News*, Vermaseren published an article about its pendant beneath the Santa Prisca, thus intensifying publicity for the Dutch excavations (Vermaseren, 1955; *cf.* Vermaseren and Van Essen, 1955–6 and Vermaseren, 1957). Once the campaigns were finished and the site was handed back to the Italian authorities, other media were also mobilised to keep the general public in The Netherlands informed. In 1959, Vermaseren authored a short, popularising book on the Mithras cult, which proved to be very successful and was translated into various languages (Vermaseren, 1959; translated as Vermaseren, 1963). In the winter of 1960–61, the Dutch cinematographer Anton van Munster, then a student at Rome's film school, took a series of somewhat romanticised, ethereal photographs of the site of Santa Prisca (see Figure 4.7); the next year, a Dutch TV crew came to Rome for a short broadcast (Heres, 1989: 87).[9] The scientific results of the excavations were eventually published in 1965, on the occasion of which Vermaseren once more reiterated his claim that 'the *Mithraeum* under and behind S. Prisca on the Aventine is without doubt the most important sanctuary of the Persian god in Rome; it may be the most interesting *Mithraeum* in the world' (Vermaseren and Van Essen, 1965: ix).

But Vermaseren was not yet satisfied. In December 1962, he had requested permission, again with the institutional backing of Van Essen, to start a new excavation in the garden south of the church of Santa Prisca. The permission was duly granted by the archaeological service in Rome, and a concurrent application for funding from ZWO was also successful. The new campaign started in 1964 and lasted for three years, but in this case resulted in few significant finds and only one short publication by Vermaseren some ten years later (Vermaseren, 1975; *cf.* Kruijer, Hilbrants, Pelgrom and Taviani, 2018). His evident disillusionment is understandable, for the excavations did not yield what he and his team expected. But another explanation for the failure of that second campaign at Santa Prisca is that the academic and political network which had so strongly favoured the first campaign in the 1950s had largely disappeared by the mid-1960s. Van Essen himself had died quite suddenly in 1963; director Poelhekke left the Netherlands Institute two years later for a professorship in Nijmegen. Ambassador Boon had already left Rome in 1958, and in The Hague, Cals left the Ministry for Culture after eleven years in 1963. With the academic and political winds changing, the Netherlands Institute in Rome came increasingly under threat, and was almost shut down by the government in the early 1970s. The favourable conditions for a successful Dutch archaeological excavation in Rome had passed. It might be seen as a sign of the times that the museum at Santa Prisca eventually had to close after a burglary, while the frescoes in the *mithraeum* progressively deteriorated – to Vermaseren's lifelong dismay. The bad state of the site in a way symbolised the ultimate fate of the Santa Prisca campaign. After decades of relative neglect, only recently have new efforts been made on behalf of the archaeological service to revitalise the site and its historical legacy.

Conclusion

The success (and eventual failure) of the first Dutch excavation in Italy give remarkable insight into the dynamics of archaeological practice in an international setting. Trained as Classicists, Van Essen and Vermaseren were dilettanti, archaeologically speaking, but nonetheless they succeeded in starting and carrying out a difficult excavation campaign in the highly complex stratigraphy of Rome. One of the factors in their success was the informal networks and therefore smooth collaboration they could establish with Italian specialists. This collaboration could materialise thanks to the spirit of international cooperation in Italy in the post-war years, which opened up Italian archaeology to foreigners after decades of *de facto* exclusion. At the same time, the academic and diplomatic networks of the Netherlands Institute in Rome, which extended to the Italian archaeological service, policy-makers in The Hague, as well as private donors, church authorities and the occasional journalist, created a favourable setting in which the Santa Prisca excavation could thrive. Within these networks, a crucial role was played by seemingly secondary and informal connections, such as with ambassador Boon or ZWO board member Van Wagenvoort, who worked hard behind the scenes to generate political, academic and financial momentum for Dutch archaeology to take wing in Rome. The history of archaeology, then, is not only a history of the big men involved in large-scale excavations, but also of the lesser known, personal connections and non-archaeological characters populating the networks that surround archaeological practice.

In the case of the Santa Prisca excavation, the result of the mobilisation of these networks was not only a successful excavation campaign as such, but also, if not especially, a successful campaign for private-public fund-raising and valorisation. The Santa Prisca excavation in a way embodied the gradual development of state-funded research in the 1950s, first approaching private capital to enable a publicly funded project over a longer term. In a period in which there is increasing uncertainty about the availability of state funding for academic research, the history of the Santa Prisca excavation might serve as an example of successful private-public partnerships. Moreover, the Santa Prisca team also understood the need to disseminate the results of their campaign to a general audience through a range of media, not least by establishing an on-site museum. That strategy of valorisation, shared by the Italian archaeological service, can equally serve as inspiration for today's archaeologists. Yet what is also striking about the Santa Prisca excavation is that the rationale for starting and publicising the campaign was very much couched in nationalist terms, as the fulfilment of a Dutch cultural claim on Rome and Western civilisation. Despite the internationalist climate after the Second World War, such a national appropriation of antiquity and archaeology remained *en vogue* long after the fall of Fascism. The resulting paradox of nationalist internationalism proved difficult to maintain, at least in the case of Santa Prisca: before long, the *mithraeum* and its museum were closed to the public, and the pioneering Dutch excavation in Italy became largely forgotten. Its spirit lived on only in the private sphere of Vermaseren's home in Amsterdam. Far away from the Eternal City, Vermaseren kept alive his fascination by constructing, in his own living room, a full-size Roman *mithraeum*.

Notes

1 The chapter is one of the outcomes of the Santa Prisca Project, developed in the autumn of 2015 at the Royal Netherlands Institute in Rome (KNIR). Following the initiative of the Soprintendenza Speciale per i Beni Archeologici di Roma (SSBAR) to map and make public all archaeological remains on the Aventine hill, the KNIR and the SSBAR decided to join forces to study and make accessible the information concerning the Santa Prisca excavations. The Santa Prisca Project was directed by Jeremia Pelgrom (KNIR) and Miriam Taviani (SSBAR). We are particularly grateful to Luigia Attilia (SSBAR) for her advice on the SSBAR archive (ASSAR), and we would also like thank Lennart Kruijer, Jord Hilbrants and Janet Mente (KNIR). For more info, see http://romearcheomedia.fub.it/aventino/, accessed 22/06/16.
2 Vermaseren's personal archive of the excavations was made available to the KNIR by his friend and colleague Joop Derksen (1938–2018); it can be consulted in the KNIR's library, Rome. The general archive of Vermaseren was recently obtained by the Allard Pierson Museum, Amsterdam. ASSAR is located in Palazzo Altemps, Rome; most of the documents relating to the first phase of the Santa Prisca excavation are in busta 417, fascicolo 8, vecchia segnatura 9/27 ('Roma. S. Prisca. Lavori di scavo e restauro eseguiti dall'Ist. Storico Olandese (1953–59)'). Complementary archives that were consulted for this research include the archive of the Associazione Internazionale di Archæologia Classica (AIAC) at Palazzo Altemps, Rome; the Archivio Centrale dello Stato, Rome; the archive of NWO/ZWO in the Nationaal Archief, The Hague; and the archive of M.N. van Lansdorp, who worked as architectural assistant to Vermaseren, in the Stadsarchief, Amsterdam.
3 On Dutch archaeology outside The Netherlands more generally, see Linde, 2012a and Dries, Slappendel and Linde, 2010. For the contemporary and somewhat comparable context in Greece, see Wagemakers, 2015.
4 In the fall of 1906 Leopold started his career as a correspondent in Rome for the *Algemeen Handelsblad*. He published extensive reports in the publication series of the KNIR, *Mededelingen van het Nederlands Instituut te Rome* (*MNIR*), but also reached a wider public with his weekly column 'Uit de leerschool van de spade' in *Nieuwe Rotterdamsche Courant* (1923–34).
5 Constant Willem Lunsingh Scheurleer (1881–1941) came from a rich banker's family. He had a great interest in Greek ancient culture and was a passionate collector of ancient Greek artefacts. In 1924, he opened his Archeologisch Museum at Carnegielaan 12 in The Hague. After the museum closed, the archaeological collection of Scheurleer was bought by the Allard Pierson Museum, Amsterdam.

6 One of the initiatives of this period that strengthened the bonds in the archaeological community in the Netherlands was the periodical *Bulletin Antieke Beschaving* (now *Babesch*). Van Essen was involved in this journal from the start.
7 See also the correspondence between Cumont and Vermaseren in the library of the Academia Belgica, Rome and in the Vermaseren archive in the Allard Pierson Museum, Amsterdam.
8 Van Essen and Vermaseren hypothesised that the rooms they excavated around the *mithraeum* belonged to a private habitation of Trajan, the so-called *Privata Trajani* (their interpretation was challenged by subsequent scholars: see Sangiorgi, 1968; Salomonson, 1969). They concluded that the *mithraeum* had been created in this *domus* around 200 AD and that it was plundered two centuries later and eventually filled up with earth to serve as foundation for the church of Santa Prisca.
9 Van Munster's photographs have been published digitally by the KNIR: https://issuu.com/knirlibrary/docs/santa_prisca_catalogue_final_2/1, accessed 22/06/16).

5

A romance and a tragedy: Antonín Salač and the French School at Athens

Thea De Armond

Defined, in culture-historical fashion, as the regions occupied by the ancient Greeks and Romans, the 'Classical world' once spanned much of Europe and parts of Asia and Africa.[1] The study of the Classical world – in particular, its archaeology – has been somewhat more limited in geographical scope, or rather, its most prominent forebears tend to hail from only a few places, namely Germany, Great Britain, France and, perhaps, the United States of America (see Dyson, 2006). It is not surprising that the history of Classical archaeology maps onto geopolitics. After all, with their shared claims to universality, Classics and empire have much in common (Porter, 2006; Bradley, 2010); Classical materials – like so many other desirable goods – gravitate toward power.

Of course, Classics has never been the sole provenance of the powerful. Even the geopolitically 'marginal' have sought their share of Classical culture (see Stephens and Vasunia, 2010), to say nothing of so-called 'source' nations such as Greece and Italy (see Hamilakis, 2007; Ceserani, 2012). However, outside the geopolitical centre, Classical archaeology often traces unfamiliar pathways, unfamiliar to those for whom Classical archaeology comprises Winckelmann, Delphi and the Vienna School.

In this chapter, I follow the career of one geopolitically marginal scholar, the Czech Classicist Antonín Salač (1885–1960), focusing particularly on Salač's waxing and waning relationships with French scholars and French institutions. Czechoslovakia had little leverage in the traditional centres of Classical antiquity. But thanks to his relationship with France – a relationship rooted in French-Czechoslovak diplomatic relations – Salač, a scholar from a 'small nation', managed to insinuate himself into what was, for the most part, a conversation between Great Powers. Unfortunately, the French-Czechoslovak relationship did not withstand the Czechoslovak Communist Party's takeover; thus, the geographies of Classical archaeology in Czechoslovakia shifted.

It merits emphasising that Salač's abortive love affair with France was not solely his own; nor is his story simply a one-off attempt by the geopolitical margins to secure a piece of the Classical archaeological pie. Rather, it hints at myriad, lesser known but no less important histories of Classical archaeology. After all, Great Powers are hardly the only states that seek power.

Introducing Antonín Salač

Antonín Salač spent most of his scholarly career at Charles(-Ferdinand) University in Prague, first as a student, later as a professor of Greek and Roman antiquities. He published hundreds of monographs, articles, reviews and 'brief notes'[2] – especially these last two – on a wide range of topics, particularly epigraphy, ancient Greek and Roman religion and numismatics; later in his career, he turned toward Byzantology (see Avenarius et al., 1992; Havlíková, 1999). Salač conducted archaeological excavations in Greece, Turkey and Bulgaria, among the first excavations under the Czechoslovak flag. However, his most significant legacy may be as an 'organiser of scholarly life',[3] as founder of the still-extant Centre for Greek, Roman and Latin Studies[4] (now, the Centre for Classical Studies)[5] at the Czechoslovak Academy of Sciences,[6] and of Eirene, a society for Classical studies in the socialist countries.[7]

Salač was a distinguished scholar by any metric, but this is not why I chose to write about him. Rather, I have chosen Salač because his career is an apt venue for the exploration of geopolitics' entanglement with scholarly practice. His biography intersects with a host of geopolitical shifts: two World Wars, the establishment of an independent Czechoslovakia and the Communist Party takeover. Moreover, Salač is one of only a few Czech Classical archaeologists to have excavated abroad, a testament to his international scholarly networks.

Salač's personal archive at the Academy of Sciences of the Czech Republic[8] also makes him an apt object of study, mostly because of its compendiousness. It covers more than forty (shelf) metres of unprocessed materials. According to Jan Bouzek (personal communication, 24 March 2014) and Pavel Spunar (personal communication, 8 August 2014), these materials essentially consist of the contents of Salač's apartment. An extensive archive is a *sine qua non* for this chapter, given its preoccupation with the basic infrastructure of scholarly production

– that is, with scholarly networks beyond those attested in published scholarship.

At issue here is an idea central to the overlapping bodies of scholarship known as science and technology studies, science studies and histories of science – namely, that scholarship cannot be walled off from 'real life' (see e.g. Shapin, 1998; Livingstone, 2003, 2005). Despite scholars' pretensions to objectivity, to 'a view from nowhere' (see Nagel, 1986), knowledge production always takes place within historical, political, sociological, economic, material, etc. contexts. David N. Livingstone writes: 'Given that bodies are resolutely located in space, there are grounds for suspecting that scientific knowledge is always positioned knowledge, rationality always situated rationality, inquiry always local inquiry' (2003: 80).

Without directly engaging with science studies, for the most part, archaeologists have already begun to 'put knowledge in its place' (see Livingstone, 2003: 1–16). Since the 1990s, studies of nationalism and archaeology – essentially, 'geographies of knowledge' at the scale of the nation-state – have flourished (see e.g. Kohl and Fawcett, 1995; Díaz-Andreu and Champion, 1996; Meskell, 1998). If archaeology builds nation-states at home, so, too, might it negotiate for nation-states abroad; hence, the study of the entanglement of archaeology with cultural diplomacy (see Luke and Kersel, 2012). This chapter is a contribution to the latter area of enquiry. As we shall see, the scholarly networks Salač cultivated with the French and the French School at Athens were shaped by Czechoslovak foreign relations, both inasmuch as geopolitics governed access (particularly, under Czechoslovakia's Communist government), but also – and more significantly – because Salač was regarded by France as a cultural envoy.

Introducing Czechoslovakia

Czechoslovakia was established as an independent state on 28 October 1918, in the aftermath of the First World War. Although Czechoslovakia was meant to embody the principle of ethnic self-determination ('one nation, one state'), 'Czechoslovak' was an ethnicity to which few self-ascribed. Rather, 'Czechoslovakism' unified Czechs and Slovaks, in part, toward the aim of gerrymandering the new state's borders, so that the demographic impact of its sizeable German and Magyar minorities would be reduced. In 1921, 'Czechoslovaks' comprised 65.51 per cent of Czechoslovakia's overall population; Germans were 23.36 per cent and Magyars 5.6 per cent (Státní úřad statistický, 1924: 60).[9] Considered together, then, Czechs and Slovaks were less than two-thirds of Czechoslovakia's population; should the two groups be divided, the German minority in particular might prove a real demographic threat.

Anti-German sentiment had a prominent place in Czech nationalism.[10] The development of a Czech literary language – and its concomitant, a Czech literary culture – had occurred in opposition to the official German language of the Habsburg Empire (Agnew, 1993).[11] So, too, the emergence of Czech-language education (see Zahra, 2004), which lay at the centre of Czech nationalist struggles (Havránek, 2009: 50). And many Czech nationalist historians – most prominently, the 'Father of the Nation' František Palacký (1798–1876) – represented Czech history as one of conflict with Germans, with the Thirty Years' War as a crucial historiographical turning point.

In the wake of the First World War, Czech nationalism's anti-German undercurrents benefited Czechoslovakia. The victorious Allies – particularly France – supported the establishment of a Czechoslovak state as a buffer against German ambitions. France was the first of the Allies to recognise Czechoslovak independence and, during the interwar period, it focused considerable diplomatic efforts on the young state (see e.g. Ort and Regourd, 1994; Marès, 2004; Michel, 2004: 15; Hnilica, 2009). In the realm of 'hard' power, France maintained treaties with the signatories of the Little Entente (Czechoslovakia, Romania and the Kingdom of Serbs, Croats and Slovenes) to protect against (particularly, German and Hungarian) encroachments upon the states' sovereignty. In the realm of 'soft' power – of which France had a long and illustrious history (see Mulcahy, 2017: 33–63) – hoping to capitalise on Czechoslovak Francophilia, France established the first of its post-war French Institutes in Prague.

The French Institute in Prague[12] offered language courses and lectures in French history and culture. It maintained a French library and supported scholarly exchange between France and Czechoslovakia. The establishment of the Prague Institute was followed by the establishment of smaller institutes in Brno and Bratislava. These institutes were part of a cohort of post-war French-Czechoslovak cultural diplomatic institutions, including the Institute of Slavic Studies in Paris,[13] as well as French gymnasia in Czechoslovakia and Czech gymnasia in France.[14]

To know France was to love France, was to welcome its civilising influence. Nevertheless, interwar Czechoslovakia did not prove quite as susceptible to French influence as France had hoped. Alfred Fichelle (1889–1968), a professor at the French Institute in Prague, blamed on long-standing German influence on Czechoslovakia the country's lukewarm reception to his country's cultural diplomatic efforts, as well as the failure of the French language to penetrate the region prior to the First World War (Hnilica, 2009: 96). It is thus, perhaps, of interest that

Salač learned French while a gymnasium student;[15] he also purports to have been an avid reader of Émile Zola in his youth.[16] Perhaps Salač was a true Francophile. Certainly, he was well equipped to take advantage of France's attentiveness to Czechoslovakia.

Antonín Salač and the French School at Athens

In February 1920, Antonín Salač, newly habilitated at the Czech university in Prague, set out for Athens to 'acquaint himself with the Classical lands by *autopsia* [first-hand]'[17] before he took up his duties at the university. Shortly after arriving in Greece, Salač sought admission to the French School at Athens,[18] whose archaeological exploits he had followed as a gymnasium professor in Bohemia. In a 1915 article on the worship of Egyptian divinities on Delos, Salač had complained of French scholars' 'habit of using as-yet-unpublished epigraphic material',[19] without publishing it forthwith (1915: 401, n. 1). Admission to the French School would grant him access to those materials.

Salač's petition was supported by the recommendations of two scholars. The first was his mentor at the Prague university, the epigrapher František Groh (1863–1940). Groh, by Salač's account, cared for him 'like a father' (1940: 411)[20] during his time in Greece; the two were close until Groh's death. The other recommendation came from Ernest Denis (1849–1921), a historian of central Europe and a supporter of the struggles of Austria-Hungary's Slavs – especially, its Czechs – for nationhood. During the First World War, Denis campaigned for Czechoslovak independence alongside the Czechoslovak National Council. He was a founder of both the French Institute in Prague, which was, for a time, named for him, and the Institute of Slavic Studies in Paris. When Denis visited Prague in October 1920, the Prague-based newspaper *Národní listy* reported, 'it was as if the spirit of [František] Palacký [the aforementioned 'Father of the Nation'] hovered above us' (Hnilica, 2009: 33).

Unfortunately, the content of Denis' recommendation is not extant; along with Groh's, it persists only in a brief note to the French Ministry of Public Education and Fine Arts[21] by Charles Picard (1883–1965), the director of the French School at Athens: 'Mr A. Salač arrived in Athens in February 1920. He was, at that time, recommended to me, simultaneously, by Mr E. Denis… and Mr Frant. Groh.'[22] We have no evidence that Denis and Salač ever met. At any rate, Denis was not equipped to assess Salač's skills as a Classicist. Thus, his recommendation of Salač must have been meant to advance French-Czechoslovak relations. After all, the French School was one of the most venerable organs of French cultural diplomacy.

Like all of the foreign archaeological institutes in Greece, the French

School at Athens – the oldest of the foreign schools – was deeply and unashamedly political. In fact, when it was first founded, the French School's mandate for the propagation of French culture was stronger than its mandate for archaeological research (Valenti, 2006: 24). In this respect, it was a close relative of the newly founded French Institute in Prague.

During the First World War, the French School at Athens had operated as a centre for Allied propaganda (Valenti, 2001: 13). French School 'Athéniens' – as its members were called – populated Léon Rey's Archaeological Service of the Eastern Army,[23] effectively serving as guides for the military (Valenti, 2001: 11–12; René-Hubert, 2010). After the war, the school's new director Charles Picard – a former member of Rey's Archaeological Service – insisted 'the French School at Athens has not lost interest in its role as a propagator of French culture abroad' (quoted in Valenti, 2001: 14).[24] It was at this time that Salač, encouraged by Picard, with whom he became close friends, joined the French School's Foreign Section.[25]

The establishment of a Foreign Section – that is, a division of the French School for those who were not French citizens – had been suggested at the time of the school's foundation (Viviers, 1996: 173–4); however, the section was not established until 1900, after the German Archaeological Institute in Athens[26] began to admit foreign scholars, particularly citizens of the Triple Alliance states. The French School's Foreign Section thus emerged in 'a clearly anti-German context',[27] as an attempt to rally France and its allies against Germany and its allies (Viviers, 1996: 175). And the war's end did not dissipate French-German tensions – as France guarded against German expansion in central Europe, so, too, did the French School clash with the German School on the archaeological front (see Fittschen, 1996). Salač was thus admitted into the French School as a representative of Czechoslovakia, as an anti-German ally (or supposed to be such)[28] of France, in accordance with French foreign policy.

Foreign policy is the correct term to use here. Membership of the Foreign Section was negotiated between governments (see Viviers, 1996: 174). In fact, when the Foreign Section was first established, the French School required that foreign governments wishing to send scholars to the school sign a convention with the French government; it amended this requirement when Belgium proved the only country willing to do so. Salač himself sought aid from a former schoolmate in Czechoslovakia's Ministry of Education and National Enlightenment,[29] to hasten the processing of his application to the French School: 'The request needs to go through the Foreign Ministry to our embassy in Prague, there, again, through the Foreign Ministry to the Ministry – probably – of Education; that is a route, which makes my head spin and which could easily take, in the normal course of things, the whole year.'[30] Alfred Fichelle, of the French Institute in Prague, also facilitated Salač's application to the Foreign Section (Hnilica, 2009: 109). Clearly, then, even without a convention with the French government, admission to the French School was a political affair.

Throughout its history, the Foreign Section of the French School at Athens has been dominated by Belgians (see École française d'Athènes, 2014 [2017]). As of November 2017, nearly half of the 111 *membres étrangers* it has hosted have been Belgian. The Swiss and Dutch – of whom there have been nineteen and twelve *membres étrangers*, respectively – also make a good showing. Only one citizen of Czechoslovakia or its successor states – Salač – has ever been admitted to the school.

During Picard's tenure as director of the school (between 1919 and 1925), its Foreign Section was uniquely diverse (that is, with respect to nationality). *Membres étrangers* hailed from Sweden, Denmark, Russia, Poland, Romania, Belgium, the Netherlands, Switzerland and, of course, Czechoslovakia. No other director in the history of the French School has presided over so diverse a Foreign Section as did Picard; Pierre Roussel's subsequent directorship saw the Foreign Section dominated by Belgium and the Netherlands once more. This diversity was in line with French cultural diplomatic policy, broadly speaking – like Czechoslovakia, Romania and Poland saw the post-First World War establishment of French Institutes (in 1922 and 1924, respectively).

Picard patently regarded Foreign Section members as cultural envoys for France. In a report to the French Ministry of Education, he claimed that foreign members 'have contributed, in their countries, to [France's] winning over public opinion.'[31] Certainly, Salač declared his 'devotion to France'[32] in his application to the French School.

Salač's decision to join the French School at Athens thus made sense from the standpoint of interwar diplomacy – the Francophilic Salač could be counted on to advocate for France; France's support of Salač might be favourably regarded by the Czechoslovak public;[33] for Salač himself, French School affiliation meant access – but it was, in fact, unprecedented. A number of Salač's gymnasium and university professors, including František Groh, had travelled to Greece with the support of scholarships provided by the Austrian government (Frolíková, 1987). When they had chosen to affiliate themselves with the foreign schools that dominated – or even had a stranglehold upon – Greek archaeology, they generally chose either the Austrian or German Archaeological Institute. After all, as Habsburg subjects, they were fluent in German.

Over the course of his first trip to Greece, Salač joined the French School's projects on Delos and Thasos, as well as in Greek Macedonia and Delphi. According to Picard, the French School's encouragement of its foreign members' participation in School projects was a means of cultivating good will in foreign members' home countries, particularly in contrast to other foreign schools, the policies of which, he claimed, were less open. This policy also expanded the French School's skilled workforce. Salač would eventually contribute to the publication of the Delphic inscriptions; likewise, with the aid of a young archaeology student in Prague, Libuše Jansová (1904–96) – later, a prominent prehistorian – Salač would produce a catalogue of Thasian amphora stamps.[34] Despite his wishes,[35] the catalogue was never published, though it did serve as the foundation for François and Anne-Marie Bon's later catalogue of Thasian amphora stamps (1957).

Antonín Salač and Greece

Salač never intended his relationship with the French School to define his career. His aim in traveling to Greece and applying to the French School had been to secure training and resources, so that he might lead 'Czechoslovak' excavations in Greece.[36] Thus, during his first trip to Greece, Salač cultivated relationships with Greek colleagues as well as representatives of the French School – Athenians as well as 'Athéniens'. Bolstered by Greece and Czechoslovakia's shared status as 'small nations' – rich in culture, if not numbers (see Masaryk, 1916; Beneš, 1925) – Salač had sought to acquaint himself with 'the new Greece'. To his former schoolfellow Otakar Sommer (1885–1940), the aforementioned contact in the Czechoslovak Ministry of Education, he wrote 'so far, after the war, few people have come to Greece who were not already known here; I, as the member of a small nation, spoke with the Greeks differently from the members of large nations. They recognised that I sincerely like ancient Greek culture and that I have a lively sympathy for the new Greece.'[37]

Salač distributed pro-Czechoslovak propaganda to the Greek government (Salač, 1920) and negotiated business deals between Greek merchants and Czech manufacturers (specifically between Greek medical professionals and Heinrich Hoffmann, a Czechoslovak glass – and glass eye – manufacturer).[38] He produced a series of columns about Greek culture for the Prague-based newspaper *Národní listy*. He unsuccessfully lobbied for the establishment of a lectureship in modern Greek – that is, Katharevousa – at the Prague university.[39] No doubt, Salač's prominence in Greece as a representative of Czechoslovakia was attributable, in part, to the lack of a Czechoslovak embassy (which would be established in 1922). But it also amounted to an attempt to insinuate himself into the good graces of Greece's political and business classes, for the sake of future research.

During that first trip to Greece, Salač also became acquainted with some of the luminaries of the Greek archaeological establishment, in particular, with the sometime director of the Greek Archaeological Service Konstantinos Kourouniotis (1872–1945). Salač seems to have approached Kourouniotis with plans for an Athenian epigraphic archive and journal, both of which he hoped to organise and administer.[40] Unfortunately, these projects were cut short, owing to turmoil in the Greek government. As we shall see, perhaps ironically, Salač's relationships with the French School would prove far more useful in facilitating research in Greece – at least, at the administrative level – than would his relationships with Greeks.

'Czechoslovak' excavations

Salač returned from his first trip to Greece in the autumn of 1921. He kept up contact with his French School colleagues over the following year, hosting his collaborator on the Delphic inscriptions Georges Daux (1899–1988) in Prague in February 1922.[41] Salač and Picard discussed the possibility of 'Czechoslovak' excavations; Picard pressed Salač to consider an expedition to the Sanctuary of the Great Gods on the Greek island Samothraki, a vast temple complex that had been explored by Austrian-, French- and Greek-led excavations in the nineteenth and early twentieth centuries: '[Y]ou would be given authorisation without difficulty and <u>I would help you financially</u>... there would be, doubtless, important discoveries to be made there. Have no illusions about the value of new excavations; hardly any of the first order remain, while the questions on Samothrace are very important.'[42]

On Picard's recommendation, then, Salač chose Samothraki. At first, he attempted to leverage Greek connections to access the site. Salač sought advice as to how he might secure an excavation permit from the archaeologist Efstratios (Stratis) Pelekidis (1880–1958), of the Macedonian Ephorate of Antiquities, whom he may have met while working with the French School in Greek Macedonia.[43] As instructed by Pelekidis, Salač then applied to the Greek Ministry of Education for permissions to excavate on Samothraki.[44] He did so in the name of the recently established, Prague-based State Archaeological Institute,[45] of which the division for 'antique' excavations appears to have been established, in large measure, with the aim of supporting Salač's excavations in Greece.

Unfortunately, the Greek Ministry of Education reported to Salač, only foreign schools residing in Greece could be granted excavation permissions.[46] Thus, Salač was compelled to turn to the French School, with the request that the French apply for the Samothraki excavation rights on behalf of the Czechoslovak government: 'In thus becoming our intermediary with the Greek government, the French School would render a great service to our institute and to the development of Greek archaeology in our land, a service which would attest to… the amicable relationship of our country with the great and powerful French civilisation.'[47] The French School readily complied. So, the 1923 French-Czechoslovak excavations on Samothraki became a largely 'French' enterprise. The excavations were – per Picard's promise – largely funded by the French School; the Samothraki excavation permit was secured in the name of the French School; Salač was even supplied with a young 'Athénien' to assist him, Fernand Chapouthier (1899–1953), a recent inductee into the school. Picard characterised the French School's support of Salač's Samothraki excavations as 'scientific liberalism'.[48] To the French Ministry of Education, he wrote, '[a] good future is expected from this enterprise, which, in principle, renews that of Asine (1922).'[49] Picard was referring to the excavations of Salač's Foreign Section colleague, the Swedish archaeologist Axel W. Persson (1888–1951), at Asine.

In the *Národní listy* Salač reported: '[T]hough we had very little money at our disposal' – a persistent motif in Salač's research – 'we obtained satisfactory results' (Salač, 1926a: 5).[50] But rather than returning to Samothraki the following year, Salač determined 'to consider a more extensive and entirely independent enterprise' (Salač, 1926a: 5)[51] – presumably, one in which he would not be compelled to rely on French money or institutions. He had already seen that such an enterprise was not possible in Greece. Accordingly, he applied himself to the western coast of Asia Minor.

In the summer of 1924, Salač set out for Turkey, with the aim of locating a site to excavate the following year. Czechoslovakia (that is, its predecessor in Austria-Hungary) had long-standing ties with Turkey (that is, its predecessor in the Ottoman Empire). Austrian sugar, production of which centred in Bohemia, represented 74 per cent of the Ottoman Empire's sugar exports (Novák, 2006: 206); Salač would secure funding for the expedition from a consortium of Czechoslovak sugar manufacturers.[52] Likewise, Austria's centre of fez production, with a majority of its exports to the Ottoman Empire, lay in Strakonice in southern Bohemia (see Purkhart, 2010). Still more significant than these economic ties was the fact that it was legally possible to secure an excavation permit in Turkey without an affiliation with a foreign archaeological institute.

Accordingly, France's influence on Salač's subsequent excavations was relatively limited. Apparently, Salač initially requested that France transfer to him its permissions for one of its excavations in Asia Minor – presumably, Teos or Aphrodisias (Bouzek, 1980: 22). When this was denied to him, at Picard's recommendation, Salač determined to excavate at Aeolian Kyme, not far from İzmir.[53] The Kyme excavations were Czechoslovak-funded and Czechoslovak-led. Salač was assisted on the project by a Czech architecture student from the Technical University in Prague, Jan Nepomucký (1895–1948). He liaised with the Czechoslovak embassy in İzmir during his excavations, becoming close to its consul Emil Kubelka. Except for the selection of the site itself and Salač's publication of partial results in the *Bulletin de correspondance hellénique* (Salač, 1927), the excavations were very much a 'Czechoslovak' affair.

If independent 'Czechoslovak' excavations were at issue, that Salač attempted to return to Kyme before going back to Samothraki is no surprise. Moreover, the Kyme expedition appears to have received wider popular coverage than had the Samothraki expedition (e.g. *New York Times*, 1925; Salač, 1926b, 1926c). Previous explorations of Kyme, by contrast with Samothraki, had been relatively limited, so Kyme had the sheen of 'new excavations'. But persistent unrest in western Turkey curtailed Salač's plans. In the summer of 1927, Salač attempted to return to Kyme. He travelled to İzmir, only to learn that the western coast of Turkey had been declared a military zone, and excavations that year would be impossible.[54] Having already secured funding for the planned Kyme expedition, Salač hastily sent a telegram to the French School at Athens, asking if he might return to Samothraki that summer.[55] As a result, Salač and Nepomucký conducted a second series of limited excavations on Samothraki.

The 1927 excavations on Samothraki would be Salač's last collaboration in the field with the French School. In the following years, he shifted his focus to the Slavic Balkans, where pan-Slavic networks preceded him (see Curta, 2013), and the French School had little to offer. In establishing the antique division at the State Archaeological Institute in Prague, its director Lubor Niederle (1865–1944) had foreseen this eventuality, observing that, if Czech archaeologists were unsuccessful in the well-trodden centres of Classical antiquity – that is, in Greece, Italy and Asia Minor – they might turn to 'some of the Slavic parts of the Balkans, where a great deal of rewarding work still awaits the Classical archaeologist and where a Czech worker would be welcomed and supported.'[56]

In summary, then, Salač's archaeological research began in one of the centres of Classical archaeology, Greece, where he was only able to work as an affiliate of the French School, an affiliation he was able to secure, mostly thanks to France's interwar foreign policy. He proceeded thence to Turkey – relatively speaking, a centre of Classical archaeology, but somewhat less central than Greece geographically (that is, *vis-à-vis* Europe), and with a less intensive history of archaeological exploitation and, concomitantly, less stringent cultural heritage laws; there, until the tumultuous early years of the Turkish Republic curtailed his research, Salač was able to work with minimal French input. By the time Salač began to work in the Slavic Balkans – in Bulgaria and the Kingdom of Yugoslavia, still more 'marginal', still less regulated – he no longer needed France.

We have no evidence that Salač was in any way hemmed in by the French School's oversight – Picard's 'scientific liberalism' appears to have built up skills and resources among its beneficiaries, without exacting anything more onerous than Francophilic gratitude from them. Nevertheless, French-Czechoslovak excavations – hyphenated excavations – limited the recognition that might accrue to Czechoslovakia. And recognition, for Salač and for Czechoslovakia, was one of the central aims of Classical archaeological excavations. Salač's story ought to be a familiar one – national aggrandisement via archaeological excavation, what Suzanne Marchand, in the German case, calls 'spiritual "propaganda"' (1996: 245). It only seems unfamiliar because we are unaccustomed to considering the motives and the means by which the geopolitical margins – 'small nations' – might seek global recognition. On his excavations and during his travels abroad, Salač devoted a great deal of energy to acquainting the world with Czechoslovakia. On Samothraki, he reported in the *Národní listy*, 'I made sure that, always and everywhere, it was known of me that I am a Czech, even if I had to again and again indicate where my country was' (1923: 9).[57]

The path from benevolent French oversight to independent archaeological excavations – and an archaeological institute – has been traversed by a number of states, which once had foreign members at the French School. The Swedish Institute at Athens, for one, was established shortly after the Second World War (you will recall that Picard invoked the excavations of Sweden's Axel Persson, Salač's colleague in the Foreign Section, as a model for Salač's Samothraki excavations). Belgium, too, the only country to sign a convention with France to facilitate its sending students to the French School (see above), now has an archaeological institute in Athens.

At the end of the 1920s, Czechoslovakia seemed to be investigating the possibility of entrenching itself on Greek soil. In 1929, on returning from Prague, where he had been invited to give a lecture, the French historian Jérôme Carcopino (1881–1970) informed the French School that Czechoslovakia sought to sign a convention with France to facilitate its sending students to the school.[58] And in February of the following year, Czechoslovakia's president Tomáš Garrigue Masaryk (1850–1937) created a fund for the establishment of a Czechoslovak institute in Athens; the fund was to be administered by Groh and Salač.[59] Neither the convention nor the fund bore fruit. As far as we can tell, the

Great Depression intervened, and then the Second World War and the Communist Party takeover. Today, there is no Czech or Slovak (and there never was a Czechoslovak) archaeological institute in Athens.

'I do not cry, I work': 1938–48

For much of the 1930s, Salač's correspondence with his French School colleagues was irregular but warm. In the latter half of the decade, he deepened his acquaintance with Prague's French community, particularly employees of French diplomatic institutions. Salač befriended the librarian of the French Institute in Prague, Madeleine Vokoun-David (1902–??), supporting her bid to become a lecturer at the Prague university;[60] he struck up an acquaintance with the French ambassador Léopold Victor de Lacroix (1878–1948) and his family, particularly his wife Mary Ann and his youngest daughter Marie (later, Rist; 1912–96); Marie, nearly thirty years Salač's junior, was the object of his (unrequited or, perhaps, Platonic) affections.[61] In 1937, Salač was admitted into the French Legion of Honour.[62] Why Salač was thus honoured only after he had essentially ceased working with the French School is unclear. Perhaps the award was intended to re-enlist Salač as an intermediary for the French state, in view of a rising Germany.

In September 1938, the predominantly German-speaking regions of Bohemia and Moravia, known as the Sudetenland, were ceded to Adolf Hitler. Present at the meeting in Munich that wrought this state of affairs were representatives of Italy, Germany, Great Britain and France; Czechoslovakia's representatives were confined to their hotel room. Ironically, the 'Munich betrayal',[63] which saw anti-French rioting in the streets of Prague, heralded a renewal of relations between Salač and his French School colleagues. The now former French School director Charles Picard deplored the Germans' annexation of the Sudetenland, but 'here, people like me are like Cassandra'.[64] To Picard, Salač responded 'I do not cry, I work… What I feel for your country is not hate, but rather – pardon the cruel word – *pity*. Poor France!'[65]

For six years, as the Second World War raged, Salač lost contact with his network of French School colleagues. Salač again heard from Picard in September 1945, by which time the war had been over for several months. Nevertheless, Czechoslovakia's postal service had not yet been restored, so Jacqueline Mazon, daughter of the director of the Paris-based Institute of Slavic Studies André Mazon (1881–1967), brought Picard's letter with her to Prague, where she would be teaching at the French Institute.[66] André Mazon was Ernest Denis' successor in the sphere of French-Czechoslovak relations.

For some time, despite the 'shadow of Munich',[67] it seemed that France and Czechoslovakia might resume interwar relations (Olšáková, 2007: 729). France attempted to restore its relationship with Czechoslovakia, in part, by decorating Czech citizens with honours (perhaps the same grounds for Salač's earlier admission into the Legion of Honour); shortly after the war, it issued so many as to stir up controversy (Hnilica, 2009: 135, n. 63). Salač was included among the ranks of the honored; he received an honorary doctorate from the University of Dijon, where his French School collaborator Georges Daux worked.

Salač accepted the doctorate on the occasion of a trip to France for the centennial of the French School at Athens.[68] At the invitation of André Mazon, Salač delivered two lectures at the Sorbonne. The cultural-diplomatic tenor of this invitation merits emphasising – like his predecessor Denis, Mazon was not a Classicist but, rather, a Slavicist and a French-Czechoslovak cultural emissary. In Paris, Salač also attended a ceremony at which the Czechoslovak ambassador to France awarded the Order of the White Lion to Albert Pauphilet (1884–1948), a former professor at the French Institute in Prague, as well as several French journalists. While in Dijon, Salač delivered a lecture on Latin inscriptions at Prague Castle, which segued into a brief discussion 'of Charles University's founder [Charles IV] and French-Czech relations during his time'[69] – obviously, framed as a medieval precedent to contemporary relations between the two nations.

The trip to France rang with the echoes of First Republic Czechoslovakia. Salač had been reunited with his French School colleagues; the pre-eminent French-Czechoslovak cultural diplomatic institutions – the French Institute in Prague, the Institute of Slavic Studies in Paris – and their representatives – Albert Pauphilet, André Mazon – had been honored and involved in the festivities. But, once again, the renewal was brief.

The break-up: 1948

The French Institute in Prague had been closed for most of the Second World War. It reopened in June 1945, but it did not stay open for long. In February 1948, twelve non-Communist ministers resigned from the Czechoslovak government in protest against the Communist Interior Minister Václav Nosek's management of the police force. They had hoped that the Czechoslovak President Edvard Beneš would not accept their resignation. Their gamble did not pay off – in poor health and, perhaps, fearing a civil war, Beneš did accept it. In a matter of days, a new Communist government was formed.

French-Czechoslovak relations rapidly declined thereafter. Heliodor Píka (1897–1949), the head of Czechoslovakia's French Alliance,[70] a pre-First World War establishment aimed at the promotion of French language and culture, was arrested and executed in a show trial, convicted of espionage for Great Britain (see Olšáková, 2008). Píka's conviction substantiated further repression of France's representatives in Czechoslovakia. The French Alliance was dissolved. The French gymnasium in Prague was dissolved, and its professors were expelled from the country. Finally, Czechoslovakia's French Institutes – the Institutes in Brno and Bratislava and, at last, the Institute Ernest Denis in Prague – were closed (see Olšáková, 2007: 732–43; Hnilica, 2009: 140–50).

Salač's relationships with his French colleagues declined in tandem with the declining status of the French in Czechoslovakia. In a curriculum vitae dating to the second half of the 1950s, Salač claimed to have severed all ties with foreign nationals after 1945.[71] As we have seen, this was not true – in fact, 1945 marked the post-Second World War renewal of Salač's ties with foreign colleagues. Salač had a strong incentive to conceal evidence of French contacts from his archives. The trial of Píka and, later, the trial and imprisonment of Salač's colleague Jindřich Čadík (1891–1979) – who, according to Jan Bouzek, was framed by way of his friendship with the French ambassador to Czechoslovakia (Bouzek, 2012)[72] – revealed these ties to be dangerous. However, by the beginning of the 1950s, if Salač's personal archives are to be believed, he was no longer regularly corresponding with his French colleagues.

There is a poignant coda to these ruptured relations. In February 1956, three years after the untimely death of Fernand Chapouthier, his co-director on the first Samothraki excavations – and perhaps not coincidentally, following the death of Czechoslovakia's 'little Stalin' Klement Gottwald – Salač sent a note and a book (of uncertain title) to Chapouthier's wife, Odette. To Salač, Odette wrote 'I leafed through it with a great deal of melancholy, thinking about the pleasure that my husband would have had at looking through it.'[73] It perhaps merits reminding the reader that Salač's relationships with his French colleagues were not merely professional.

A romance and a tragedy

In the preceding pages, I have laid out Salač's career-long relationship with France, particularly, with the French School at Athens. His relationship with the French School ebbed and flowed, according to Czechoslovakia's relationship with France. First, as a rising docent in post-First World War Czechoslovakia, bolstered by French-Czechoslovak mutual appreciation, with its roots in shared anti-German sentiment, Salač joined the French School's Foreign Section. With the aid of the French School, Salač led excavations in Samothraki and Turkey. Throughout, he aimed for independent, Czechoslovak excavations, which he nearly achieved at Kyme. Salač's network of relationship with French colleagues declined over the 1930s and was nearly ruptured by the Second World War; it was, at last, well and truly broken – as far as we can tell from his archive – by the 1948 Communist takeover. Thus, the relationships that brought Salač to prominence apparently came to naught. His post-1948 networks – the Czechoslovak Academy of Sciences, Eirene – are the associations that lasted.

How, then, does a geopolitically 'marginal' scholar make his way in the pre-eminent 'centre' of archaeologies? In Salač's case, he did so by leveraging the cultural-diplomatic policies of the geopolitical centre to his benefit – by insinuating himself into its network. Notwithstanding Salač's scholarly prowess, he likely would have found himself unable to entrench himself in Classical archaeology without the support of the French School. Thus, if Salač's relationship with the French was a casualty of historical contingency – that is, of the Communist Party takeover – it also benefited from historical contingency – France's interwar cultivation of Czechoslovakia.

Of course, Salač's relationship with France was not solely a matter of historical contingency. It derived from the geographies that shaped – and continue to shape – Classics. Classical archaeology might cross borders – inasmuch as the Roman Empire did – but it does not transcend them. Certain parts of the Classical world were more or less accessible to citizens of Czechoslovakia than they were to citizens of France. Accordingly, Classical archaeology – its ambit and its history – takes a different form in Czechoslovakia than it does in France.

We must not be too hasty to consign Salač's relationship with France to the dustbin of history. Shortly before the Communist takeover, Salač's student, the archaeologist, Jiří Frel (1923–2006) travelled to France, as Salač had to Greece, bearing what must have been a somewhat ambivalent letter of introduction (if the draft in Salač's Academy of Sciences archive is anything to go by)[74] to Albert Pauphilet, formerly of the French Institute in Prague but, lately, head of the École normale supérieure in Paris. Frel – a dubious figure to many 'western' archaeologists but foundational in Czech archaeology – would bring his students to Paris, foremost among them Jan Bouzek and Jan Bažant, and he would be buried there, at Père Lachaise Cemetery.

Notes

This chapter is dedicated to my father, M. Keith De Armond. Thanks to Anna Hofmanová De Armond for checking my Czech transcriptions. Thanks to Angeliki Anagnostopoulo for the translation of modern Greek archival materials.

Thanks to Mark Pyzyk, Julia Roberts, Kathleen Sheppard, Ulf Hansson and Jonathan R. Trigg for comments on drafts. All errors are mine.

1 All institutions, titles and quotations not originally in English will be translated in the text. Footnotes will contain institutions, etc. in their original language. Unless otherwise stated, all translations are mine. Secondary texts will be cited in-text. All archival materials will be cited in the footnotes.
2 'Drobné zprávy', published in the preeminent Czech philological journal, *Listy filologické*.
3 'Organizátor vědeckého života' is a Czech phrase, with no exact English equivalent.
4 Kabinet pro studia řecká, římská a latinská.
5 Kabinet pro klasická studia.
6 Československá akademie věd.
7 Salač also edited the first issue of the journal *Eirene*. At the time, it was not clear that *Eirene* would be issued regularly; its second issue did not appear until after Salač's death (see Frolíková and Oliva, 2013).
8 Akademie věd České republiky.
9 The remaining 5.5% were made up of 'Russians' – including Russians, Ukrainians and Ruthenians (3.45%) – Jews (1.35%), Poles (0.57%) and 'others' (0.19%) (Státní úřad statistický, 1924: 60). Nationality remained a slippery affair in Czechoslovakia for some time (see Sayer, 1996; Zahra, 2004).
10 Czech nationalism – particularly, the Czech national revival – has been the topic of numerous scholarly works (e.g. Kočí, 1978; Macura, 1983; Hroch, 1985; Agnew, 1993; Gellner, 1994; Holy, 1996).
11 See Macura, 1983 for an excellent analysis of the significance of language to Czech nationalism.
12 Francouzský institut v Praze/L'Institut français de Prague.
13 L'Institut d'études slaves.
14 Particularly given the centrality of language to both French and Czechoslovak national identity, such foreign-language gymnasia were inherently cultural diplomatic institutions. Gymnasia are the equivalent of high schools.
15 Masaryk Institute and Archives of the Academy of Sciences of the Czech Republic, Prague (hereafter, MÚA AV ČR), Antonín Salač, inventory 410, box 29, results of Antonín Salač's *maturita*, 12 July 1904. As of January 2019, this archive has not been processed.
16 MÚA AV ČR, Antonín Salač, inventory 410, box 28, fragment of memoir by Antonín Salač, n.d.
17 '[P]oznati z autopsie půdu klasickou'. Archives of Charles University, Philosophical Faculty 1882–1966 (1970), Prague (hereafter, AUK FF UK), inventory 637, box 55, letter from František Groh to the Professoriate of the Philosophical Faculty, 1919. The Greek term *autopsia* means 'seeing for oneself'; by using it, Groh intends to invoke ancient historians' use of the term.
18 École française d'Athènes.
19 '[Z]vyk, užívati materiálu epigrafického, dosud nepublikovaného'.

20 '[J]ako otec'. Given that Salač's father died when Salač was young, this statement is particularly poignant.
21 Ministère de l'instruction publique et des beaux arts.
22 'Il m'a été, à cette date, recommandé simultanément par M. E. Denis... et par M. Le Prof. Frant. Groh.' French National Archives, Paris (hereafter F), F/17/13599, letter from Charles Picard to the Ministry of Public Education and Fine Arts, 17 July 1920.
23 Service archéologique de l'Armée d'Orient.
24 '[L]'École française d'Athènes ne s'est désintéressée de son rôle de propagatrice de la culture française à l'étranger.'
25 Section étrangère.
26 Deutsches Archäologisches Institut.
27 '[D]ans un contexte nettement anti-allemand'.
28 It merits noting that Salač does not display any particularly anti-German sentiments. He maintains strong ties with German colleagues at Charles University – especially with the epigrapher Heinrich Swoboda (1856–1926) – and, even after the Second World War, aids and communicates with German colleagues ousted from the university.
29 Ministerstvo školství a národní osvěty.
30 'Žádost musí přes ministerstvo zahraničí k našemu vyslanectví v Paříži, tam zase přes ministerstvo zahraničí na ministerstvo asi školství, to je cesta, ze které se mi točí hlava a která by mohla trvati při normálním běhu krásně celý rok.' MÚA AV ČR, Otakar Sommer, letter from Antonín Salač to Otakar Sommer, 5 April 1920.
31 '[O]nt contribué, dans leurs pays, à nous gagner la faveur de l'opinion publique'. F, F/17/13598, report on work from October 1920 to October 1921 from Charles Picard to the Ministry of Public Education and Fine Arts, 30 October 1921.
32 '[S]es sentiments dévoués envers la France'. F, F/17/13599, letter from Charles Picard to the Ministry of Public Education and Fine Arts, 17 July 1920.
33 Of course, the effectiveness of 'soft power' is notoriously difficult to assess (see Mulcahy, 2017: 34–7).
34 MÚA AV ČR, Antonín Salač, inventory 410, box 21, draft of letter from Antonín Salač to Charles Picard, n.d.; MÚA AV ČR, Antonín Salač, inventory 410, box 36, letter from Charles Picard to Antonín Salač, 23 October 1922; MÚA AV ČR, Antonín Salač, inventory 410, box 36, letter from Charles Picard to Antonín Salač, 26 April 1923.
35 MÚA AV ČR, Antonín Salač, inventory 410, box 7, draft of letter from Antonín Salač to Günther Klaffenbach, 23 September 1956.
36 That is, Czechoslovak-led. On Salač's excavations – as, of course, on the majority of archaeological excavations at this time – diggers were locals. See Quirke, 2010 for a much-needed reminder of the significance of diggers to archaeological expeditions.
37 '[P]o válce přišlo dosud do Řecka málo lidí, kteří by tu nebyli známi; já jako člen malého národa hovořil jsem s Řeky jinak, než s nimi mluví členové národů velikých. Poznali na mně že mám starou kulturu řeckou upřímně rád a pro nové Řecko že mám živé porozumění.' MÚA AV ČR,

Otakar Sommer, letter from Antonín Salač to Otakar Sommer, 21 October 1920.

38 See e.g. MÚA AV ČR, Antonín Salač, inventory 410, box 18, draft of letter from Antonín Salač to Henry [Heinrich] Hoffmann, 31 March 1920; MÚA AV ČR, Antonín Salač, inventory 410, box 18, draft of letter from Antonín Salač to Henry Hoffmann, 26 April 1920; MÚA AV ČR, Antonín Salač, inventory 410, box 18, copy of letter from Antonín Salač to Henry Hoffmann, 29 April 1920.

39 Salač would help establish modern Greek studies in Prague, thirty years later.

40 AUK FF UK, inventory 637, box 55: letter from K. Kurouniotes (Konstantinos Kourouniotis) to the deacon of the Philosophical Faculty of Charles University, 5 June 1920.

41 See e.g. MÚA AV ČR, Antonín Salač, inventory 410, box 36, postcard from Georges Daux to Antonín Salač, 20 February 1922.

42 '[O]n vous donnerait sans difficultés l'autorisation, et je vous aiderais pécuniairement... il y aurait sans doute des découvertes capitales à faire. Notez qu'il ne faut pas s'illusionner sur la valeur des chantiers nouveaux; il n'y en a plus guère qui soient de 1er ordre; tandis qu'à Samothrace, les questions sont très importants.' Underlining present in original. MÚA AV ČR, Antonín Salač, inventory 410, box 36, letter from Charles Picard to Antonín Salač, 14 July 1922.

43 MÚA AV ČR, Antonín Salač, inventory 410, box 36, letter from Efstratios (Stratis) Pelekidis to Antonín Salač, 1922.

44 MÚA AV ČR, Antonín Salač, inventory 410, box 14, draft of statement about Samothrace excavation permissions, written by Antonín Salač but signed by Lubor Niederle, 16 April 1923.

45 Státní archeologický ústav.

46 MÚA AV ČR, Antonín Salač, inventory 410, box 14, draft of statement about Samothrace excavation permissions, written by Antonín Salač but signed by Lubor Niederle, 16 April 1923.

47 'En devenant ainsi notre intermédiaire auprès du gouvernement grec, l'École *Française* rendrait un grand service à notre Institut et au développement d'archéologie grecque dans notre pays, un service; qui témoignerait... [les] relations amicales de notre patrie à la grande et puissante civilisation française.' MÚA AV ČR, Antonín Salač, inventory 410, box 14, draft of statement about Samothrace excavation permissions, written by Antonín Salač but signed by Lubor Niederle, 16 April 1923.

48 '[L]ibéralisme scientifique'. F, F/17/13598, report on work from November 1922 to August 1923 from Charles Picard to the Ministry of Public Education and Fine Arts, August 1923.

49 'Un bon avenir est attendu de cette entreprise qui, en son principe, renouvelle celle d'Asine (1922).' F, F/17/13598, report on work from November 1922 to August 1923 from Charles Picard to the Ministry of Public Education and Fine Arts, August 1923.

50 '[P]řes to, že jsme měli k disposici obnos velmi malý, dosáhli jsme slušných výsledků'.

51 'abych pomýšlel na podnik rozsáhlejší a úplně samostatný'.

52 MÚA AV ČR, Antonín Salač, inventory 410, box 14, budget for the Kyme expedition, 1925.
53 F, F/17/13598, report from Charles Picard to the Ministry of Public Education and Fine Arts, 30 September 1925.
54 MÚA AV ČR, Antonín Salač, inventory 410, box 36, letter from the Greek Consultate in İzmir, 28 July 1927.
55 MÚA AV ČR, Antonín Salač, inventory 410, box 36, letter from R. Demangel to Salač, 27 July 1927.
56 '[V] některých slovanských částech Balkánu, kde na klassického archeologa čeká ještě mnoho vděčné práce a kde český pracovník byl by rád viděn a podporován'. Quoted in MÚA AV ČR, Antonín Salač, inventory 410, box 9, memorial of the organisation of foreign research of the State Archaeological Institute, n.d.
57 'Snažil jsem se, aby se o mně vždy a všude vědělo, že jsem Čech, i když jsem musil znovu a znovu vykládati, kde leží ma vlást.'
58 F, F/17/13599, letter from Pierre Roussel to ?, 24 June 1929.
59 MÚA AV ČR, Antonín Salač, inventory 410, box 14, letter from Antonín Salač to the Ministry of Finance, 15 December 1948. In 1948, the fund appears to have been included in an assessment of Salač's personal property. In this letter to the Ministry of Finance, Salač argues against that asssessment, relating the history of the fund and his role as its manager.
60 Vokoun-David was a philosopher, Orientalist and translator – as well as a librarian – whose *Debate about Writing and Hieroglyphs in the 17th and 18th Centuries and the Application of the Idea of Decipherment to Dead Writings* (*Le Débat sur les écritures et l'hieroglyphe aux XVIIe et XVIIIe siècles et l'application de la notion de déchiffrement aux écritures mortes*) inspired Jacques Derrida's *Of Grammatology* (*De la grammatologie*).
61 MÚA AV ČR, Antonín Salač, inventory 410, box 6, draft of letter from Antonín Salač to Marie de Lacroix, n.d.
62 Légion d'honneur.
63 'Mnichovská zrada'.
64 '[L]es gens comme moi, ici, jouent les Cassandre.' MÚA AV ČR, Antonín Salač, inventory 410, box 21, letter from Charles and Gilbert-Charles Picard to Antonín Salač, 10 November 1938.
65 '[J]e ne pleure pas, je travaille... Ce que je sens pour votre patrie, ce n'est pas une haine, mais – pardonnez moi le mot cruel – plutôt *une pitié*. Pauvre France!' MÚA AV ČR, Antonín Salač, inventory 410, box 21, draft of letter from Antonín Salač to (?Charles Picard), 1938.
66 MÚA AV ČR, Antonín Salač, inventory 410, box 4, letter from Charles Picard to Antonín Salač, 2 August 1945.
67 'Stín Mnichova'.
68 For a detailed account of Salač's trip to France, see AUK, FF UK, inventory 637, box 55, report on Antonín Salač's November 1947 visit to France, 17 November 1947.
69 '[O] zakladateli university Karlovy a stycích česko-francouzských za jeho doby'. AUK, FF UK, inventory 637, box 55, report on Antonín Salač's November 1947 visit to France, 17 November 1947.
70 Alliance française.

71 MÚA AV ČR, Antonín Salač, inventory 410, box 29, CV of Antonín Salač, n.d. (post-1955).
72 See also Morávková and Řehoř, 2012 for an extended account of Čadík's trial, for which Bouzek is a key source.
73 'Je l'ai feuilleté avec beaucoup de mélancolie en pensant au plaisir que mon mari aurait eu à le parcourir.' MÚA AV ČR, Antonín Salač, inventory 410, box 1, letter from Odette Chapouthier to Antonín Salač, 4 February 1956.
74 MÚA AV ČR, Antonín Salač, inventory 410, box 1, draft of letter of introduction for Jiří Frel by Salač, n.d.

6

Geographies of networks and knowledge production: the case of Oscar Montelius and Italy

Anna Gustavsson

In this chapter, I aim to highlight the potential of thinking geographically when studying networks and the production of archaeological knowledge, by considering the contacts in Italy of the Swedish archaeologist Oscar Montelius (1843–1921, see Figure 6.2) and his work on Italian prehistory.[1]

Oscar Montelius was a pioneer of prehistoric archaeology from the late nineteenth century onwards. He is mainly known for his work on typology and chronology. His *Om tidsbestämning inom Bronsåldern med särskildt afseende på Skandinavien* (1885), is still frequently cited.[2] Montelius held positions at the National Museum of Stockholm, eventually became the director of the Swedish National Heritage board and was involved in numerous excavations in Sweden. He also travelled all over Europe, more than most scholars at the time, to compile and study archaeological finds, and is still one of few who have managed to study and publish on such a vast number of archaeological artefacts. His wife Agda Montelius (1850–1920) was deeply involved in her husband's work, and accompanied him on several research trips. Montelius became affiliated with the National Museum as a young scholar in the mid-1860s and was awarded a doctorate in history, since archaeology was not yet an established academic discipline. His dissertation, entitled 'Remains from the Iron Age of Scandinavia' (published as Montelius, 1869), was an overview of current research on how Iron Age culture spread from Egypt, via Greece, Rome and Hallstatt to Scandinavia. From an early stage, an important characteristic of his research method was to gather as complete a collection of examples for each object type as possible

6.1 Oscar Montelius' study and desk in his home in Stockholm. (Date unknown). Ref: Riksantikvarieämbetet arkiv, ATA, Montelius-Reuterskiölds samling, FIV a1 Fotografier.

(Baudou, 2012a: 181). When Montelius studied the prehistory of Europe archaeological sites and finds in Italy were of fundamental importance for understanding and constructing chronologies for the Bronze and Iron Ages. He was the first scholar to produce a synthesis of the Italian Bronze and Iron Ages, *La Civilisation Primitive en Italie depuis l'introduction de métaux*,[3] which was published in five volumes (1895–1910) covering northern and central Italy.

The past twenty years or so have seen a development of geographical thinking among historians of science, which is signified by the acknowledgement that spatiality – spaces and places of different kinds – influence the production of knowledge in a variety of ways. Such ideas were a reaction to earlier notions that (good) science is placeless (Livingstone, 2003: 1–5; Naylor, 2005: 2). To me, adapting geographical perspectives when writing histories of archaeology offers the possibility to add new perspectives and methods to a field (archaeology and its history) where these ideas are relatively new. As pointed out by Withers and Livingstone (2011: 1), a great number of aspects and themes of science can be analysed by thinking of knowledge production in geographic terms,[4] and they claim furthermore that:

> ... the geographies of nineteenth-century science present a particularly fertile intellectual terrain for inquiries of this sort: science was being given disciplinary shape in this period as perhaps not before; science became in this period a public good with a variety of audiences and staging places; science's disciplinary emergence was evident in certain discursive procedures and methods that helped define 'the field' in question; particular forms of dissemination, be they lectures, specialist journals, or instrumental procedures, helped give science a public and professional credibility not readily enjoyed in earlier periods (Withers and Livingstone, 2011: 4–5).

In my opinion, this is applicable to the development of the archaeological discipline and therefore for the study of the production of archaeological knowledge. Archaeological knowledge is created in a variety of places. The location of research activities affects both how science is conducted, and the result of the science; science is not independent of local settings (see Livingstone, 2003: 1–3 for a discussion of situated knowledge production). Naylor states that while science is always carried out in 'circumscribed localities', it also is always situated in a general context. With this in mind, I would like to stress that the history of archaeology and the production of knowledge can, and needs to, be studied in the light of regionalism, nationalism, trans-nationalism and internationalism. I agree with Livingstone (2003: xi) that scientific findings 'are both local and global; they are both particular and universal; they are both provincial and transcendental'. This makes it particularly interesting and useful to study a scholar's work done outside their own country. Oscar Montelius was a Swedish scholar partly financed by the Swedish state, who spent many years studying finds and sites in and from different Italian regions, meeting a variety of scholars, collectors and professionals involved in heritage and culture politics. He obviously needed to learn how to navigate local and regional Italian customs, while being in the midst of geographical and political changes, regional struggles and competitions for influence following the recent unification of Italy in 1861. (Rome did not become the capital until 1870–71.)

Here I will provide examples of Oscar Montelius' contacts within the Italy-based scholastic community, focusing on the geographical areas Bologna/Emilia-Romagna and Rome, during the last decades of the nineteenth century. Despite the importance of Italy for Montelius' work, little detailed examination has been made of his specific whereabouts and personal connections within this context. It soon becomes clear that an enterprise such as *La Civilisation Primitive* cannot be analysed as a one-person job. To scrutinise the work process and production of archaeological knowledge that took place in a variety of geographical spaces and places also means studying and understanding the scholarly networks of the time. I believe that a fruitful methodological approach includes thinking in terms of co-production, collaboration and contribution, but also controversy. Another important aspect in my opinion would be to consult and combine a large variety of sources. By doing these things, I believe it is possible to gain a wider understanding of how archaeological knowledge is created, and to learn more about the premises and factors that affected individual scholars' work at the end of the nineteenth century. Here, I analyse the creation of knowledge by studying the travels of the Montelius couple and the work related to Oscar Montelius' massive work on Italy, *La Civilisation Primitive*, which I treat as a great project as well as a publication. Focusing on an analysis of the project is a way to relate the different source material, and to help identify the dynamics of these networks and their impact on Montelius' research.

Previous research on Oscar Montelius and his work

Previous research on Oscar Montelius' work has often focused on important but quite specific aspects such as typology and the accuracy and legacy of, in particular, his Bronze Age chronology (see, for example, Gräslund, 1974). However, little focus has been put on investigating the conditions under which artefacts were studied, and how the international scholarly community developed in detail. In addition, many Swedish publications and articles on the history of archaeology are not published in English. There are few monographs about Oscar Montelius. The archaeologist Hanna Rydh published a biography, the first, in 1937. Evert Baudou published a second, extensive biography of Montelius (2012a), in which the relationship between his work, origins and his personality are discussed. Baudou is the leading Montelius scholar to date. He has discussed and developed ideas and a methodological approach to biographical studies of individual scholars that is based on the theories of Ludwik Fleck.[5] These have been thoroughly discussed elsewhere in this volume, and will not be further explained here. Important features in Baudou's work are the relationships between a scholar's life and research, as well as relationships and interactions between an individual and the collective. He presented his method (2012b) in comparing the parallel work and scholarly progress of Montelius and Hildebrand, two colleagues and friends. In the case of Montelius, the discussion includes his role in scholarly groups, or thought-collectives (Baudou, 2012b: 194). Baudou's research forms a solid base for further studies as he has carefully mapped the life, work and personality of Montelius. He underlines Montelius' friendly, outgoing personality and exceptional social skills, but also characterises him as being a very independent individualist, with strong self-esteem (Baudou, 2012a: 378–9). This image of Montelius seems accurate in many aspects. The most recent publication on Montelius is Patrik Nordström's (2014) book, *Arkeologin och livet*. Nordström's work is a portrait of Agda and Oscar Montelius and contains a collection of 902 letters, transcribed from the 2,500 letters sent between them during the period 1870–1907 that have been preserved. In a short introduction, Nordström outlines the basic facts about the couple and the archive material. These studies of Montelius have been based mainly on Swedish archive material.

The sources: digging deeper in the archives

The rich Montelius archive is located at the Riksantikvarieämbetet, Antikvarisk-topografiska arkivet (ATA) in Stockholm. There are actually two collections: the Oscar Montelius archive and the Montelius–Reuterskiöld collection, the latter of which also includes the archive of Agda Montelius. Drawing on diverse types of material, but also from different archives, offers the possibility of finding cross-references and discrepancies in the material, which add new dimensions to the research. The archive material used for this chapter includes the diaries of Agda Montelius and private correspondence as well as 'work material' from the Oscar Montelius files.

I would like to underline the importance of placing the work of Montelius and other scholars in a wider geographical, historical and political context. To add further depth to an analysis of both individual and general perspectives on a scholars' work, I suggest always consulting and combining a variety of sources and archives. Studying a famous person who has already been a subject of research can be complex, though. The archaeologist Oscar Moro Abadía (2013) provides a good example of how multiple sources can be used, and how a national narrative can be challenged by changing focus, in his article 'Thinking about the concept of archive: reflections on the historiography of Altamira'. He questions the traditional story of the discovery of the Altamira cave paintings, and the view of the scholar Juan de Vilanova as being a radical pioneer, and presents an alternative interpretation of the history of the discovery and Vilanova's role. Moro Abadía succeeds in doing this by compiling and comparing archival sources in an innovative way. I believe, as Moro Abadía argues, that it is important to broaden the analysis to be able to present alternative and new narratives of well-known persons like Montelius, and in the process recognise more unnoticed factors, actors and the conditions of their time in which research was carried out. Moro Abadía does not use geographical thinking explicitly in his short article, but it would be possible, and could be fruitful, to discuss for example intellectual and political geographies that influenced national narratives about Altamira and Vilanova.

La Civilisation Primitive

> Outside Italy I have only one debt, but it is a very deep one. It is to the late Prof. Oscar Montelius, whose literary executors – The Swedish Academy – have generously allowed me to reproduce a number of drawings, some of them original, from the three-volume atlas of plates entitled, *La Civilisation Primitive en Italie*. This exhaustive work should be the incessant companion of all who are studying the Italian prehistoric cultures. (Randall-MacIver, 1924: vi)

This quotation comes from the preface to David Randall-MacIver's work, *Villanovans and Early Etruscans*. Published in 1924, it was one of the first attempts to make a synthesis of the Iron Age in Italy, building directly on the work of Oscar Montelius. Since then, Montelius' publication on Italy has frequently been (and still is) cited internationally, while seeming to have sunk into obscurity among Scandinavian archaeologists. As mentioned, Montelius was the first to publish an extensive work on prehistoric Italy. He never completed the volumes on South Italy, Sicily and Sardinia, which were supposed to end *La Civilisation primitive*. Nevertheless, the publication consists of five large volumes on the north and central Italian peninsula.

During the nineteenth century, the means for travelling and communication changed drastically, at least for those who could afford them, owing to technical and infrastructural advances in Europe. Oscar Montelius went all over Europe to study collections and artefacts. He often travelled together with his wife Agda, who became very engaged in his work. He was involved in debates on Mediterranean research from an early stage. Adopting a comparative research method, Montelius tried to see as many collections and finds as he could, and to date, his work actually remains the most extensive study of the Italian material. In the archive in Stockholm there are a vast number of notes, sketches and working materials prepared by the couple. Agda Montelius made most of them, and she kept a fairly detailed travel journal during their trips.

Why did Montelius carry out such a large study? Previous research suggests that early in his work on establishing a final typological chronology for the Bronze and Iron Ages of northern and Central Europe, he found it necessary to turn to the Mediterranean area. The combination of finds, literary sources and references to Egyptian material could help with dating artefacts (Gierow, 1995: 67). Previous research implies/

suggests that his master plan was quite clear from the beginning, but this was probably not the case. Greece would have been a good starting place for Montelius' work for many reasons, but Italy is closer to Scandinavia than Greece, and it was more likely that direct links to Scandinavia would be found there. The most important reason for choosing Italy was above all the archaeological activity and new research that had begun at the time (Gierow, 1995: 68; Guidi, 2008: 113; Sassatelli, 2015: 9). For a long time it was also easier, for logistical and political reasons, to travel to and within Italy than to Greece.

The beginning of his travelling and his networks (1870s and 1880s)

Scientific activities are dependent on, but also generate, places and spaces for research activities (see for example Naylor, 2005: 3). These include both physical/material and intellectual/abstract spaces. Probably the most important spaces or venues for the international development of the archaeological discipline and the formation of networks in the 1870s and 1880s (apart from museums) were academic congresses. In 1865, the subtheme *paletnologia* (prehistory) was introduced at an Anthropological and Geological Congress in La Spezia. It was there decided that an 'International congress on Anthropology and Prehistoric Archaeology' (CIAAP) should take place regularly. At this time, both cultural exchange between countries and scholarly interaction seems to have been quite limited among Scandinavian scholars, at least when it came to learning of regional and local discoveries, collections and finds in Europe (Baudou, 2012b: 184–5). The CIAAP congresses, and other international meetings, were about to change all that. An important question at the time was whether the past should be organised according to a three-age system, and how different ages/periods should be placed and related. In 1869, the congress convened in Copenhagen, where the young Oscar Montelius made his big entrance onto the international scene. This was probably where he first met or heard of several of the scholars who would be a part of his professional network throughout his career. Among the Italians were Luigi Pigorini (1842–1925), at the time the director of the museum in Parma, Giovanni Capellini (1833–1922), Professor at the University of Bologna and Count Giovanni Gozzadini (1810–87) who was involved in the first discoveries at and excavations of the Villanova 'culture' near Bologna. It was at the next Congress, in Bologna in 1871, that the Bronze and Iron Ages gained serious attention for the first time.[6] The congress was scheduled to take place in 1870 but had to be postponed because of the Franco-Prussian War (Sommer, 2009: 18–19). International conflicts have constantly changed the political map, something that heavily affected scientific activities and interactions during the period of my study. As Livingstone puts it, 'imagined geographies have real consequences' (Livingstone, 2003: 8). In the meantime, Agda and Oscar Montelius had married, and the trip to the Bologna congress (see Figure 6.2) was part of their honeymoon (Bokholm, 2000: 49).

The archaeological congresses reflected international politics and national aspirations in the mid- and late-nineteenth century, but of course also individual research interests. The fast development of prehistoric archaeology, especially in the North Italian regions, might be looked upon as a reaction against the earlier focus on Roman and Classical periods and part of the struggle for independence as well as for the unification of Italy. In this melting pot of archaeological discoveries, political ideas and the formation of nation-states, the first broad generation of modern professional archaeologists met and formed networks. Evert Baudou (2012b: 194) has identified and suggested at least four thought-collectives of different character in which Montelius took part. Two of them were international; one is defined as South Scandinavian, the fourth as a group of Bronze Age scholars (non-specified), born in the 1840s. Fleck's (1979: 38–51) theory of thought-collectives can be perceived as distinct and accessible, and provides a base for further studies of the international networks which sprung from the 1869 congress in Copenhagen.

The congress publications shed light on what scientific questions were considered important and which scholars and dignitaries were present. Private letters between the scholars can tell us about what happened between congresses and how trans-national networks took form. The Italians were not surprisingly in the majority at the Bologna congress and it is not possible to know how many of them Montelius actually talked to. However, it is possible to compare the list of letter senders in the Montelius archive with the listed delegates to study how it corresponds. The North Italian archaeologist Edoardo Brizio (1846–1907) was not listed as a delegate at the Bologna congress, but he was engaged in preparing an archaeological exhibition for the conference. Montelius took great interest in the display, which presented material from every Italian region. Despite not being recorded, Brizio is present in a group photo of the delegates and it is very plausible that Montelius met him and talked to him at the congress. Brizio had a strong interest in both the Classical and prehistoric periods as well as the origins of different people; he had an impact on the future work of Montelius and some of his Scandinavian colleagues. The congress was of great importance for Montelius, and the congress events were the start of more extensive debates concerning the Iron Age. With the discoveries near Bologna

6.2 Oscar Montelius, portrait from the Bologna Congress 1871. Ref: Riksantikvarieämbetet arkiv, ATA, Montelius-Reuterskiölds samling, FIV a1 Fotografier. Copyright © Riksantikvarieämbetet. All rights reserved and permission to use the figure must be obtained from the copyright holder.

came the possibility for Montelius and others to study the material in context, connecting typology with stratigraphy. Montelius would later return to Bologna on several occasions and was included in another type of geographical space, an international scientific and intellectual circle, arranged by the Countess Maria Teresa Gozzadini, the wife of Count Giovanni Gozzadini (Vitali, 1984: 224; Guidi, 2008: 110).

The earliest preserved letter between Brizio and Montelius is from 1882. Brizio had been working in Pompeii, Rome and other places, but from the 1880s he was mainly active in Bologna and the surrounding region. He held different positions within archaeological and museum administration, as well as at the university, and was therefore an important person to know for a foreign scholar like Montelius. In addition to Brizio and Pigorini, the scholars Giuseppe Bellucci (1844–1921) in Perugia and Pompeo Castelfranco (1843–1921) in Milan are among those who stand out at this stage as having been extremely helpful. Even though Bellucci was based in Perugia, he was apparently of assistance when the Montelius couple visited Bologna and other cities north of Rome. Montelius maintained contact with many Italian scholars over long periods. In some cases, the correspondence was relatively frequent, as with Pigorini. Over 60 letters between 1874 and 1919 are present in the archive, which proves regular contact in between meetings in person. In other cases, as with Bellucci, the letters are fewer (about a dozen between 1876 and 1902) although the meeting in person would have been very important to Montelius. Bellucci was listed as a delegate at the Bologna congress, while Castelfranco was not. Further investigation is needed into when and how Montelius met and interacted with each of these important Italian scholars.

Unfortunately, there are no diary notes by Agda Montelius from the Bologna congress. Her first accounts of their journeys to Italy are from 1876. Oscar Montelius had been awarded the generous Letterstedska travel grant, which allowed them both to travel for almost eighteen months in total, during 1876 and 1878–9. Judging from Agda's diary, the couple were extremely busy and productive. In addition to sightseeing and getting to know Rome, Bologna and other cities, they made new connections and visited well-known people and scholars. They also began their studies and documentation of archaeological finds in museums and private collections. In the early days of December 1876 Agda made several notes about meetings in Rome with people like Augusto Castellani (1829–1914), Wolfgang Helbig (1839–1915) and Luigi Pigorini. Pigorini had transferred to Rome and from the mid-1870s was both the director of the national prehistoric museum and the professor of prehistoric archaeology at the Università di Roma (today known as La Sapienza). In mentioning Helbig, it might be important to point out that the early scholars travelled to Rome long before most of the foreign institutes and academies had been founded. The first was the German archaeological institute (Deutsches Archäologisches Institut, DAI), established in 1888. DAI was a continuation of the Istituto di corrispondenza archeologica, a 'German' (Prussian) initiative from 1829. The institute was therefore an obvious platform for foreign scholars. Helbig held an important position at the institute from 1865 and kept it for two decades. With help from the staff at the institute, the scholars were introduced to the Roman scholarly community.[7]

After a few days in Rome in December 1876, the couple went to Naples where Agda worked on drawings of finds in the National Museum while Oscar searched for and bought *fibulae* and other small objects from the local art dealers.[8] It should come as no surprise that he, like many others, was buying artefacts, but it is worth noting, especially as he complained about collections and objects lacking context, for example at the museum in Tarquinia. From the diary notes, it is clear that the couple travelled back and forth in Italy to study finds and collections.

Judging by the registered correspondence in the Montelius archive at ATA as well as his publications, it was from the early to mid-1880s that his work on Italy's prehistory became more organised. The Montelius couple made several trips to Italy during the 1880s. In 1881 Oscar and Agda returned to continue the collection of data. Agda's diary tells that on 13 June 1881 Oscar Montelius met Pigorini at the Rome prehistoric museum for many hours and studied the museum collection, which was followed by a visit to Wolfgang Helbig. Judging from the diary, Helbig was helpful and gave advice concerning a visit to Orvieto that the couple had planned for the next day. Helbig was not always so accommodating, as we shall see. In the Italian context, political power struggles and possible tensions between regionalism and nationalism, Rome and other cities, seem to have had important impacts on the means to carry out research and produce knowledge. What did it mean to be a foreign researcher in this setting? Montelius was in many aspects a complete outsider, being Swedish and not living permanently in Italy. Being local but also German, like Helbig, could certainly generate tensions as well. When is a person an insider or an outsider in a scholarly community? (See for example Livingstone, 2003: 19–20 on aspects of inclusion and exclusion.)

If one wishes to build on the concepts of thought-collectives discussed by Fleck (1979: 38–51) and the research of Baudou (2012a and b) to understand and exemplify more of the international context of the research Montelius conducted, one way could be to look for other thought-collectives within the local Italian context. However, I

6.3 Examples of original illustrations and sketches sent to Oscar Montelius by his Italian colleagues. Ref: Riksantikvarieämbetet arkiv, ATA, Oscar Montelius arkiv, F1b Arbetsmaterial F1b vol 148. Copyright © Riksantikvarieämbetet. All rights reserved and permission to use the figure must be obtained from the copyright holder.

find it necessary to incorporate the study of networks or clusters in a wider and more inclusive framework. Here, thinking geographically about science and knowledge production provides useful tools. After mapping a number of places (such as sites or museums) and people (their location and professional/social position) found in the travel notes and letters, it is possible to start suggesting different networks or clusters in which Montelius took part, which have to be considered. The networks might overlap each other, or be revised later on, depending on the connections between the actors. To further explore and understand the work process, mechanisms and knowledge produced in the work of Oscar Montelius on Italy, I suggest thinking in terms of different degrees of contribution, collaboration and co-production, but also controversy. For some Italian colleagues, and a few Scandinavian scholars/archaeologists, who sent written information or sketches from different locations in Italy (see for example Figure 6.3), it is quite difficult to 'assess' the individual and exact impact/significance of their contribution to *La Civilisation Primitive* (and other work by Montelius). Co-production might therefore be too strong a term in most cases. We can definitely state, however, that collaboration is a valid description for several cases where Montelius received written information and illustrations of finds and sites from colleagues on an informal and friendly basis, without being charged. In other cases, he had to formally order and pay for sketches and other items. Ongoing analysis of the archive material indicates that the former, more informal ways of cooperation were more common in northern Italy, than Rome. For the 1870s and 1880s, I suggest the following groups (in addition to the special example of his wife Agda) to be considered in relation to the work of Montelius: scholars based in northern Italy, scholars based in Rome, and art dealers and collectors.

In the case of Agda Montelius there is no doubt the term co-production is the most appropriate. She was involved in the travel planning, took the more detailed notes of the two, compiled informative notes and made sketches of finds and also of their day-to-day experiences (see Figure 6.4).When she did not accompany her husband on his journeys, she very often acted as his stand-in at the museum in Stockholm, performing a variety of tasks and enabling him to take time-off (Bokholm, 2000: 51–2). One particular example of co-production is an article about Sardinia. It was published under Oscar's name in the journal *Ymer* (Montelius, 1883: 31–5). However, when consulting the diary of Agda from their trip to the island; one discovers that parts of the article are a re-writing of her diary (from 1879).

Consolidation and competition (1890s to early-1900s)

During the 1890s, the Italian state formulated stricter judicial regulations for heritage management and excavations, especially in relation to foreign scholars. Montelius made another study trip to Italy in 1895, the year that the first volume of *La Civilisation primitive* was published. He returned in 1898. Even if there were few new finds to see, Montelius was still missing certain drawings and considered it necessary to revisit collections to make additional comparative studies before concluding his work. Competition and rivalry had existed all along, but seems to have increased during this period. When tracing the visits of Montelius it becomes obvious that political and scholarly rivalry had hardened. Even though Montelius had gained an increased scholarly status internationally, it became more difficult for him to get access to certain collections in Rome. Conflict, which would also affect a foreign scholar like Montelius, stemmed from enmity between the first director of the Museo Nazionale Romano,[9] Felice Barnabei (1842–1922) and Wolfgang Helbig. In short, the conflict between Barnabei and Helbig

6.4 A Sardinian man in Cagliari. Sketch from the diary of Agda Montelius, 1879. Ref: Riksantikvarieämbetet arkiv, ATA, Montelius-Reuterskiölds samling, F2B.1a.

concerned new finds from the Faliscan culture, a Latin-speaking 'people' in northern Latium. Helbig accused Barnabei of unprofessional excavation methods, wrongly documenting the contexts in which finds had been made, and displaying them poorly.

Barnabei was appointed general director of the state cultural authority in 1895. He was still very much involved in the museum work, but Edoardo Gatti (1875–1928) was introduced as operative director at the museum in 1899. However, the archive material suggests he might have functioned as the director long before that. In a letter to Agda from spring 1898, Montelius wrote about the difficulties of locating both Barnabei and Gatti to ask for permission to study the collections. In the letter, Montelius referred to the former as having been promoted to general director, and the latter as being the operative director of the museum.[10] After finally finding Barnabei at the museum a day later, Montelius was told that he could not make any notes or sketches while studying the collection. He had to write a formal application to Barnabei. Montelius is mostly described as a gentle person by others, but referred to his encounter with Barnabei as 'lively'. Barnabei wrote about the same event in his diary, where he referred to Montelius as bursting in like 'a hyena', demanding to see this and that (Barnabei and Delpino, 1991: 245). When Montelius mentioned the problem to Helbig at a dinner the same evening, the latter considered the whole thing to be a scandal. Despite the problematic start, the archive material shows that Barnabei and Montelius were soon on speaking terms, and that Montelius got permission to study the collections. Montelius was also among the few foreign scholars who expressed his support for Barnabei in the conflict with Helbig (Barnabei and Delpino, 1991: 254). At the end of his study visit in May 1898, Montelius decided to invite his colleagues to dinner to show his gratitude. He found it necessary to arrange two separate dinners, which clearly illustrates the cold social relations between the Italian and German archaeologists at the time. As an extra courtesy to Barnabei, Montelius invited him first, and asked him to choose the date for the dinner.[11] To succeed, a foreign scholar like Montelius had to interpret how to behave within the local and regional 'intellectual' structures. This meant, as Livingstone puts it, 'unpacking the implications and inferences that are fixed in local structures' (Livingstone, 2003: 6–7).

From time to time the sometimes rather helpful Helbig would also obstruct Montelius' investigations, for example by delaying communication with other (Italian) scholars. The importance of having 'agents', both when Montelius could not be present in Rome or Italy himself, and when navigating the networks, appears to have become greater and greater. During the 1890s, Montelius did not rely just on his Italian connections, but also on Swedish/Scandinavian agents or mediators. One such person, and a person of high political status, was the Swedish diplomat Carl Bildt (1850–1931). He was more or less based in Rome from the 1890s onwards. Bildt was interested in history generally and in promoting Swedish research. He wrote letters to scholars like Felice Barnabei and Wolfgang Helbig, and mediated and arranged meetings with the aim of enabling/facilitating for Montelius. He also kept

Montelius informed about current disputes and archaeological news in Rome:

> The enmity between Helbig and Barnabei is unfortunately fiercer than ever. Which of the two is right, I do not dare think, but I have observed that most of my acquaintances agree with Helbig. Recently, at the Forum, highly interesting excavations have taken place, and even more promising ones are imminent in the upcoming months. At the same time several 'restorations' are taking place, that I, for my part, would rather see undone. It would be lovely if you could come here soon and see the new finds. (letter from Bildt to Montelius, 16 January 1899)[12]

Another helpful person was the Swedish Classical archaeologist Sam Wide (1861–1918). In 1893, Wide received the Letterstedska travel grant and spent six months in Italy (Berg, 2016: 67). Judging from the letters he wrote to Montelius during his trip, one wonders how he had time for his own studies. He must have spent a lot of his time helping Montelius, and collaborating with Carl Bildt, to get access to information and sketches of finds that Montelius needed for *La Civilisation Primitive*.[13] Wide, being twenty years younger than Montelius and aiming for an academic career in Sweden, might have acted for career-strategic reasons. In addition to looking at excavated objects, Montelius also discussed with Wide the possibility of setting up excavation projects in Crete and elsewhere in Greece.[14] These plans were never realised by Montelius, but the correspondence between them can serve as an illustration of the international competition for 'big digs' in the Mediterranean region, fuelled by both national agendas and personal prestige.

Following the correspondence of both Bildt and Wide to Montelius, it is necessary to consider even more categories of networks or clusters in addition to the three groups mentioned above. These would include diplomats and visiting Swedish scholars, subgroups in the Roman scholarly network and the division between Italian scholars and non-Italian, especially German, Rome-based scholars.

Concluding thoughts

In this chapter, I have highlighted the potential for using a geographical approach when studying knowledge production and when writing histories of archaeology. Archaeology is itself a geographic discipline. As Simon Naylor (2005: 2) points out, it is not a finite result in itself to claim that science can be interpreted geographically. It is rather a foundation for empirical narratives. It has proven very useful so far to use Montelius' personality, and his connections with Italy, as a starting point for studying knowledge production during the early days of international prehistoric archaeology. He and his wife were able to travel more than most scholars at the time. Many of the trips were motivated by Montelius' extensive work on the prehistory of Italy, and could be realised thanks to travel grants and other funding. Montelius did study collections of Italian artefacts in other European countries, but the majority of the material was in Italy. His personal meetings with Italian scholars and dignities often made it possible for him to gain knowledge of, as well as access to, private collections and new discoveries. The mapping of research activities includes, in the case of Montelius' archaeological sites, locations of collections, of institutions and of key persons. How the latter move between geographical places, and/or within social/political strata affects the practices and premises of knowledge production.

By adding methodological tools and concepts from network studies and consulting a variety of source material, I believe that it is possible to pinpoint factors that contributed to the formation of a scholarly network, and to reconstruct the most important features of the research processes and work of Montelius and his colleagues. Analysis of archive sources has made it clear that it is of great importance to study the nationalities, geographical locations and relations between the actors, rather than 'just' the specific discoveries and archaeological data, to understand how knowledge was produced and disseminated. The extensive travelling and network building through personal meetings were necessary for Montelius' research. A notable feature is that there is a relatively high representation of Italian scholars in the private correspondence of Montelius, with many of whom he stayed in contact over long periods, regardless of where they were affiliated. He appears to have moved quite easily between the different scholarly traditions of Classical and prehistoric archaeology. He also succeeded in maintaining good relationships with scholars who were enemies, like Barnabei and Helbig.

His specific role in, and the dynamics of, different networks has to be further investigated, but it can be stated that his work was not an individual effort. It must be analysed in terms of different degrees of co-production and collaboration, as well as controversies. By thinking geographically of science and knowledge production, it is possible to identify and map different geographies/categories that affected the work process, as well as the outcome. When building on previous studies on Montelius, I would also suggest making a clear distinction between the methodological concepts of thought-collectives and different types of networks or clusters. The networks/clusters might partly overlap other networks and thought-collectives, or not do so at all, but more importantly they serve as a structure that enables (or disables) the production and dissemination of knowledge. It is possible to identify both formal and informal structures, and they too overlap each other. Just like layers on a digital map, they can be lit/switched off in a continuing analysis of the networks.

It is important to consider that the means available and practices of scientific activity must often be understood within the context of a region (Livingstone, 2003: 88; Naylor, 2005: 7). One trait so far is the suggestion of a more informal 'climate' in Northern Italy, and a more formal setting in Rome. Another is that Montelius needed to pay less often when he asked for material from northern Italy, than from Rome. What relations did really impact on the premises of knowledge production? Can we really tell whether and how a schism affected the research process and the development of the archaeological fields of research? In some cases, the answer is yes. A concrete example of a formal way to stop or delay Montelius in his work would be not granting him official permission to study a collection, or refusing to send him a requested drawing or data of a find in the name of the state authority, as Barnabei did at first. Such situations had some impact on Montelius' research and the spread of knowledge.

A case study of Oscar Montelius and his project *La civilisation primitive* clearly benefits from geographical aspects of knowledge production, but can also serve as an example of how the history of archaeology can add to the geographical understanding of scientific development, both when it comes to the conditions for the individual scholar, and for the European scholarly community in general.

As always, when scratching the surface, many new questions can be raised. In which respects were Montelius representative and unique, for his time? Why did a Scandinavian – not an Italian – publish a synthesis of Italian prehistory?

Notes

1 Thanks go to Ingrid Berg, Evert Baudou and Patrik Nordström for generously sharing their knowledge of how to navigate in the archives.
2 This book was originally published in Swedish, with a French summary. In 1986 the book was republished and translated into English with the title *Dating in the Bronze Age: with special reference to Scandinavia* (Montelius, 1986).
3 From here on shortened to *La Civilisation Primitive.*
4 In English the word 'science' traditionally refers to the natural sciences. In this chapter, I make no distinction between different disciplines. I include archaeological research in the term.
5 Fleck's ideas and concepts have been frequently used by Swedish scholars

since the 1990s, which is worth noticing. In 1997 *Genesis and development of a scientific fact* was translated into Swedish by Bengt Liliequist, who also published a dissertation (2003) on Fleck at the Department of Philosophy and linguistics at Umeå University, Sweden. At the same university, Evert Baudou was the professor in Archaeology between 1975 and 1991. These circumstances could probably provide a geographically inspired case study on thought-collectives in itself.

6 This can be established by studying the themes and content of the congress publications chronologically, and by reading contemporary travel/conference reports; see for example Hans Hildebrand (1872). *Den arkeologiska kongressen i Bologna: Berättelse.* Stockholm.

7 Wolfang Helbig left his formal position at the Istituto di corrispondenza archeologica in the mid-1880s, but stayed very much involved in archaeological research and art dealing. He and his wife, the Russian princess Nadine, held a strong social position and were often the centre of attention, inviting scholars to their home in Rome for scientific and cultural evenings. See Chapter 3 for further information.

8 Riksantikvarieämbetets arkiv, ATA, Montelius-Reuterskiölds samling, F2B 1a, Agda Montelius, diary notes, December 1876.

9 At the time, the museum had two sections, one housed in the baths of Diocletian, the other in Villa Giulia, a former papal estate a stone's throw from Piazza del Popolo, just outside the Roman city wall. At the beginning of the museum's history, the collections were arranged properly in the section in Villa Giulia, but not the other. The finds included objects from Latium and its pre-Roman inhabitants, mainly collected from current excavations. The two museums remain in the same buildings today, but the bureaucratic organisation and collections have changed over time.

10 Oscar Montelius to Agda Montelius, 13 May 1898, in Nordström, 2014: 331.

11 Oscar to Agda Montelius 24 May 1898 and 30 May 1898 in Nordström, 2014: 334, 337; Biblioteca Angelica, Rome, Fondi Barnabei 237/3, letter from Montelius to Barnabei, 24 May 1898.

12 Riksantikvarieämbetets arkiv, ATA, Oscar Montelius arkiv, E1a vol 4.

13 See for example Riksantikvarieämbetets arkiv, ATA, Oscar Montelius arkiv, E1a vol 42, letters from Wide to Montelius, 7 April and 27 July 1893.

14 Riksantikvarieämbetets arkiv, ATA, Oscar Montelius arkiv, E1a vol 42, letter from Wide to Montelius, 12 April 1901.

7

'More feared than loved': interactional strategies in late-nineteenth-century Classical archaeology: the case of Adolf Furtwängler

Ulf R. Hansson

Knowledge production in archaeology and elsewhere in academia is naturally dependent on the interaction between actors who connect, cluster and collaborate on fieldwork or other projects, and exchange information or test out new discoveries and ideas with colleagues within the various institutional and informal structures of the discipline such as university departments, professional societies, museums, congresses, workshops, journals, networks, etc. The strong social nature of these creative processes has long been acknowledged and applies to the whole field, including its so-called 'instrumental' actors. We all build on the achievements of others in our field and seek contact and exchange with colleagues working on similar material. Most of us are grateful for the opportunity to meet face to face, and we often stress the importance of collegiality and interaction for our own professional development. But not all of us are socially skilled; quite a few dread the pressure that the social arenas of the discipline generate and reproduce, while others are viewed as 'toxic' controversialists creating unwanted friction within the community. Tension and friction are constant presences, and perfectly legitimate professional disagreements that constitute a vital part of any healthy scientific or scholarly process can easily deteriorate into open conflict, even lifelong feuds, of a more personal kind that risk destabilising the dynamics of these institutional and informal structures, disrupting communication channels and forcing actors to rethink their positions and interactional modes and strategies. Much has been said about collegiality and the benefits of archaeologists coming together, but structural and interpersonal friction or conflict within the community, whether potentially constructive or mainly counterproductive, and the various effects on the dynamic processes of knowledge production and dissemination, constitute equally important aspects that have been less studied. The environments in which we operate inform, stimulate and restrict our work, speech and actions, regardless of whether they are perceived as mostly positive or negative in character (Montuori and Purser, 1995: 83; Livingstone, 2003; Bourdieu, 2004). Conflicts and disputes potentially impinge on where, why and how research is planned, conducted, presented and received. Based on a fairly well-documented but little studied case from the formative period in the modern history of Classical archaeology, this chapter explores how dynamic scholarly processes can be affected when a so-called 'key actor' in the community feels excluded, disrupts or withdraws from certain social aspects of the profession while at the same time is struggling to maintain, even reinforce his (in this case) shifting positions and strategic moves within its overlapping networks and clusters.

The professional career and scholarly production of Adolf Furtwängler (1853–1907; see Figure 7.1) constitute an interesting case of such disruptive dynamics. Focusing on Furtwängler's problematic relations and interaction with the scholarly community, of which he nevertheless saw himself as an undisputed member throughout his professional trajectory, this chapter addresses the problem of individual–collective tension in networks and knowledge production. When examining such interaction – what is being said and done by various actors as well as the reaction of their audience – the crucial importance of the physical and social spaces where all this is taking place has been acknowledged (e.g. Goffman, 1959; Livingstone, 2003; Bourdieu, 2005: 148). The social space, or 'field' to use Bourdieu's terminology, in which knowledge is generated and negotiated, is both structuring and structured by its institutions, networks, clusters and individual actors, and further regulated by specific protocols and practices recognised by its actors. A scholar's professional trajectory occurs within this 'dynamic ever-shifting relational structure of positions and unfixed boundaries' (Lipstadt, 2007: 40) and can thus be said to consist of a series of negotiated relational positions and relocations or moves that are strategic and both require and confer 'capitals' of various sorts. This is an account of one such negotiated trajectory.

Furtwängler is today fairly well-known as a pioneer of Classical archaeology. During his lifetime he was almost equally known within the scholarly community for his ill temper and propensity for polemic. In the surviving testimonies and documentation he comes across as a man for whom friction characterised much of his interaction with colleagues at institutions in Berlin and Munich, where he was active for most of his career, but also with the scholarly community at large. At the

7.1 Adolf Furtwängler (1853–1907). Deutsches Archäologisches Institut Zentralarchiv (used by kind permission). Copyright © Deutsches Archäologisches Institut Zentralarchiv. All rights reserved and permission to use the figure must be obtained from the copyright holder.

same time, the awkward work situation that resulted from this friction seems to have somehow spurred his own creativity and productivity, or perhaps these were strangely unaffected by it. A combination of personality traits, negative work experiences and strategic positioning seems to have made Furtwängler place himself outside and at times even in open conflict with sections of this community and some of its key members, and he was in turn socially isolated by many of his colleagues (Reinach, 1907b; Bissing, 1907; Hauser, 1908; Church, 1908; Furtwängler, 1965: 231*f*.). But he was never or rarely marginalised as a scholar, rather the opposite: Furtwängler in fact managed to be both 'feared and respected by all' (Reinach, 1907b) or rather, 'more feared than loved' (Perrot, 1900), and his work was mostly well received. The creative urge into which he seems to have channelled much of these perceived negative experiences and resentment resulted in a series of highly focused projects and widely influential books, several of which were later canonised as milestone publications (e.g. Furtwängler, 1890, 1893, 1900).

The popular histories of the discipline mention Furtwängler only in passing, if at all. Still, in many respects he perfectly embodies the eccentric, exceptionally gifted, restless and feverishly working but socially handicapped and temperamental 'Great Man' that their readers love to hear about. This 'lone genius' type of scholar that to some extent still prevails in the popular imagination is often perceived as someone who manages to be creative by struggling against or at least rising above the constraining forces of the field's institutions and its conforming masses. This creates the unfortunate impression that actors identified as 'instrumental' are able to produce something new and original not as *a result of* interaction with the collective, but rather *in spite of* it. In the case of such creative people, schizoid or deviant behaviour is often romanticised and even viewed as synonymous with 'genius' (Montuori and Purser, 1995: 74). Such tendencies to decontextualise individual actors who are identified as 'instrumental' are highly problematic. Furtwängler is no exception. A French colleague claimed in his obituary of Furtwängler that his work 'bore the mark of genius', and asked why we should leave it to posterity to use this word for it (Reinach, 1907b).

What perhaps makes his case somewhat peculiar is that the negative aspects of his personality and his aggressive mode of interaction had been foregrounded and famously ventilated in public during his lifetime (see especially Perrot, 1900; Gardner, 1907; Hauser, 1908; Reinach, 1928). They were woven into the dense and rather successful mythology of ill-tempered genius that was in fact created around his persona from very early on, by himself and others, and this has no doubt affected the reception of his substantive and wide-ranging contribution to the discipline. An English colleague, Percy Gardner, suggested in his obituary that 'of the many thousands of pages which he printed, perhaps not one does not contain something of value' (1907: 252). To his favourite student and later colleague Ludwig Curtius, Furtwängler's achievements and importance for the discipline of Classical archaeology were fully comparable to those of Mommsen for Roman history and Wilamowitz for Classical philology (Curtius, 1958: 224, often repeated by later authors, e.g. Calder, 1996). These were views that were shared by many of his contemporaries, who generally were greatly impressed by Furtwängler's strong dedication, grasp of data and breadth of knowledge, but who did not fail to recognise some of the flaws in his often ambitious constructs and the conclusions he drew from his study material. The scarce critical scrutiny that his work has attracted in recent years and his relative absence from modern histories of archaeology have resulted in wildly divergent assessments of him ranging from a 'largely forgotten figure' (Marchand, 2007: 252) at one extreme to 'probably the greatest archaeologist of all time' (Boardman, 2006: 20) at the other. One thing is certain, Furtwängler's influence has been considerable and is still strong, as we continue to build – at times rather uncritically – on his contribution to the discipline.

Born to middle-class parents in Catholic Freiburg, Adolf Furtwängler spent four years studying Classical philology and philosophy at Freiburg and Leipzig before turning to archaeology in Munich under the charismatic Heinrich Brunn (e.g. Curtius, (1935) 1958; Schuchhardt, 1956; Straub, 2007: 21–77; Wünsche, 2007; Hansson, 2014). Graduating at the age of 21 with a doctoral dissertation on Eros in Greek vase-painting (Furtwängler, 1874), his professional trajectory began in the late 1870s with fieldwork in Italy and Greece, where he documented museum collections and worked on material from the excavations at Mycenae and Olympia. It continued with habilitation at Bonn University under Reinhard Kekulé in 1879, followed by fifteen years as assistant curator at the Berlin Museums during their most expansive period (Furtwängler, 1965; Curtius, 1958: 215). It peaked with his appointment in 1894 to the prestigious Munich chair in archaeology and the directorship of four important museum collections in the city, and ended with his premature death in 1907 while doing fieldwork on Aegina in Greece. He was buried in Athens, and the considerable reputation he enjoyed at the time of his death is confirmed by the fact that the Greeks honoured him with a state funeral. In the three decades that he was active, Furtwängler had gained extensive experience of field archaeology, curatorial work, and higher education at leading German institutions. He had produced some twenty monographs, several of them multi-volume works, and hundreds of journal articles, encyclopaedia entries and book reviews. Towards the end of his life he was one of the best-paid archaeologists in Germany, and his extensive travelling and all of his research and field projects were liberally funded from start to finish (Zazoff, 1983: 207). His was indeed a most distinguished career. But in Furtwängler's own mind a handful of negative experiences, especially from his years in Berlin where he never felt sufficiently recognised for his work or even accepted (e.g. Reinach, 1907b; Hauser, 1908; Schuchhardt, 1956: 17), cast a deep shadow over every later success. They not only affected his interaction with colleagues and his network building, but interfered with his own research – how it was planned, carried out, and presented. More about this below.

Furtwängler was active during the formative period of the modern discipline when Classical archaeology was liberally funded by the state and enjoyed the highest esteem in Germany, both within academia and outside (e.g. Marchand, 1996; Bažant, 1993: 103). This was also the height of positivism when big, state-funded excavations like the Olympia project, in which Furtwängler participated (Furtwängler, 1890; Marchand, 2002), and Pergamon yielded not only quantities of artworks but notably great masses of less exciting bits and pieces from past human activity that had to be processed and explained to the general public. Furtwängler participated in all this with great enthusiasm as a field archaeologist, museum curator, teacher and public educator, moving from Munich to Italy and Greece, Bonn, Berlin and then back to Munich again, invariably working within high-profile structuring institutions, or in their shadow. Towards the end of his life he enjoyed a considerable international reputation as one of the great 'oracles' of the discipline, a connoisseur who was consulted on all aspects of ancient culture, even though his speciality remained sculpture, vase painting and the minor arts. But even if his wide-ranging activities have made a deep and lasting impact in several fields of study, as mentioned there is surprisingly little critical discussion of them in the discipline's official histories.

There are a number of possible reasons for this. Furtwängler's pioneering achievements in the field concerned classification rather than more spectacular discoveries and were mostly carried out early in his career within projects that he did not direct himself but where he worked in the shadow of charismatic personalities such as Ernst Curtius at Olympia and Heinrich Schliemann at Mycenae (Furtwängler and Loeschcke, 1879; Furtwängler, 1886, 1890). He then turned from publishing excavation material, which nevertheless proved to be an extremely useful experience, to re-assessing previously collected artworks in public and private collections, producing a series of detailed studies and catalogues that were often based on a thorough first-hand knowledge of large parts of the preserved *corpus* of objects (e.g. Furtwängler, 1883–7, 1885, 1893, 1896, 1900). Although important, such work is seldom recognised in conventional histories, which focus more on great discoveries in the field. Moreover, Furtwängler, who worked at the height of positivism and had a rather naïve attitude towards the potential and limits of research and knowledge (Curtius, (1935) 1958: 214; Schuchhardt,

1956: 21), did not introduce any radically new theories or methods, but instead had a rigorously systematic approach to great masses of objects that proved extremely influential within the discipline. This talent for systematisation and structure, combined with great willpower and a very good visual memory, lent Furtwängler the nickname 'the Linnaeus of archaeology' (Riezler, 1965: 9) and he was soon recognised as an international expert on a wide spectrum of materials, such as sculpture, bronzes, pottery, vase painting and engraved gems. His typologies and classification systems have tended to survive much better than his ambitious historical syntheses and the conclusions he drew, which often reveal strong cultural, racial and other prejudices. Every piece of information that Furtwängler uncovered in the field, in museums, private collections, auction houses, research libraries as well as in correspondence and conversation with colleagues was meticulously collected in a vast private archive and also mentally recorded: 'very little remained unknown to him' (Hauser, 1908: 466). In this respect, Furtwängler resembled the younger J.D. Beazley, although his data collecting had a much wider scope. Furtwängler took a great interest in photography and very early on understood how to take advantage of its full documentary potential for the study of sculpture, vase painting and gems, and its value as a teaching tool in general (Dally, 2017). He shared this interest with the Munich archaeologist and collector Paul Arndt, who for a brief period was his assistant and possessed the financial means to collect photo documentation of artworks systematically and on a large scale. Although they shared many interests, especially sculpture and gems, even Arndt fell in and out of favour and soon lost his position as assistant (Bulle in Furtwängler, 1965: 231; Curtius, 1950: 209*f*.; Moltesen, forthcoming).

Furtwängler did not like crowds and was an awkward public speaker, ill-prepared and usually talking ad hoc and in a clipped manner over impressive series of diapositives (Hauser, 1908: 468; Curtius, 1958: 215; Furtwängler, 1965: 231*f*.; Wünsche, 2007: 303). No doubt the diapositives, a great novelty back then, were part of the attraction. But they do not explain Furtwängler's extreme popularity as a speaker and the large audiences he invariably drew, which during the Munich years were rarely below a hundred, sometimes twice as many (Hauser, 1908: 468; Furtwängler, 1965: 231*f*.). His open lectures were attended by a motley crew of royalty, socialites, artists, colleagues and students, many of whom came to experience in real life this remarkable personality that they had heard so much about. A French colleague complained ironically (or sarcastically) that one had to book seats well in advance if one wanted to attend (Reinach, 1907a). Social gatherings, especially the Berlin salons, but even his wife's cultural *soirées* in Munich, invariably made him uneasy (Curtius, 1958: 215).

7.2 Adolf Furtwängler with fellow members of the Bureaux et Comité Executif at the First International Congress of Archaeology in Athens 1905. *Comptes rendus du Congrès International d'Archéologie*, I[e] session (Athènes: Imprimerie Hestia 1905, p. 147). Public domain image.

Furtwängler seldom appears in group photographs; at least, very few survive. Two are interesting in this context, for different reasons. In the first one (Figure 7.2), Furtwängler is seen posing with fellow members of the Bureaux et Comité Executif at the First International Congress of Archaeology in Athens, 1905 (CIA, 1905: 147). The photograph is exceptional in that it is one of exceedingly few instances where Furtwängler appears side by side with colleagues. This particular one includes individuals with whom he was barely on speaking terms, such as his one-time superior at the Berlin Museums, Alexander Conze, and the director of the German Archaeological Institute in Athens, Wilhelm Dörpfeld. Furtwängler is placed in the upper right-hand corner, at a convenient distance from those with whom relations were frosty. Chance meetings with colleagues that he disliked or had offended publicly often proved extremely awkward. The historian of religion Jane Harrison also belonged to this group. One of her students recorded an episode in Athens in 1901, when Dörpfeld had invited Harrison to the German Institute while Furtwängler was staying there, in all likelihood with the intention of embarrassing or discomfiting Furtwängler. 'D[*örpfeld] introduced him to J[ane]*. "You know Miss H[arrison]. I think you met in Berlin' v[.] smilingly – F[urtwängler] as stiff as a poker and looking furious. J[ane] rose in her most gracious manner and forced him to shake hands' (J. Crum, unpublished diary, 1901, p. 22, quoted in Stray, 1995: 126*f*.). According to his students, who knew him well, Furtwängler

7.3 Adolf Furtwängler with his closely knit seminar group during their 1905 excursion to Vienna. Lullies 1969, pl. 16. Public domain image.

generally had difficulty connecting with people, and as a result had very few friends apart from his much-loved wife Addy (Adelheid) (Curtius, 1958: 222). The few colleagues that he seems to have been at all close to included his early mentors, Brunn and Curtius, the Munich Latinist Ludwig Traube (Hauser, 1908: 464; A.E. Furtwängler, 2005: 16) and one or two fellow students from Munich, notably Georg Loeschcke (they too fell out in the early 1900s: Hauser, 1908: 464).

Furtwängler's relations with his students were rather different. The second group photograph of interest here (Figure 7.3) shows his closely knit seminar group during their 1905 excursion to Vienna. There are several women in the photo. Not everyone was a student of his, but Furtwängler is known to have freely admitted women to his classes and examinations long before they were officially given access to higher education in Germany (Church, 1908: 64; Lullies, 1969). He did not live to see any of them graduate, but Margret Heinemann (1883–1968), on the far left of the upper row in the photograph, was one of the first women in Germany to be awarded a PhD in archaeology, graduating in 1910 from Bonn University under Furtwängler's former friend Loeschcke (Wehgartner, 2001: 271–3). Both German students and those of other nationalities later attested that, despite his hypercriticality and contrary to what was mostly the case with his colleagues, Furtwängler was very appreciative in his teaching role and invariably valued and respected the opinions expressed in his seminar; at times he even showed real or feigned surprise when he realised that his students did not always know quite as much as he did about a certain topic (e.g. Bissing, 1907; Church, 1908; Curtius, 1958: 223*f.*; A.E. Furtwängler, 2005: 10). Students thus provided much of the face-to-face intellectual stimulus and exchange that Furtwängler needed for his work. His graduate seminar, with its shifting localities in lecture rooms, museum galleries or cast collections, became the sacred laboratory space, his *Arbeitsinstrument* (Schuchhardt, 1956: 18), where Furtwängler could be entirely at ease. Here, photographs or casts of ancient art objects were closely examined, and originals – sent there for his expert opinion – were carefully unpacked, inspected and discussed within this narrow circle. Several testimonies to this fact survive (e.g. Bulle in Furtwängler, 1965: 231*f.*). Students were also invited to Furtwängler's home in Munich or his country house on the Tegernsee. Perhaps he viewed this close-knit group as a model for what academia should ideally be like.

Unfortunately, this was late in his career, throughout which he had wrestled with paranoid tendencies and an uncontrollable bad temper. In an unpublished autobiographical sketch written at the age of 25, Furtwängler confesses that a major fault of his is that he is too rash in presuming, even taking for granted, that other people are hostile towards him or despise him, a personality trait that had plagued his father also (quoted in A.E. Furtwängler, 2005: 10). This started early. He was officially excused from secondary school to finish his *Abitur* exam from home (Schuchhardt, 1956: 7) and did his utmost to avoid military service: 'I cannot express how detestable I find this whole military business', he confessed in a letter to his teacher and mentor, Heinrich Brunn.[1] He was also absent for long periods from the graduate seminar in Munich on the grounds of illness, real or imagined – all attested by preserved correspondence.[2] After finishing his PhD, a brief attempt at teaching high-school Latin proved disastrous, and he commented that all the other teachers saw him as arrogant and aiming for something better (quoted in A.E. Furtwängler, 2005: 10). In an early letter to Brunn, he expresses his fears about his upcoming exams, as he had discovered that one of the examiners was someone with whom he was not on speaking terms (Hofter, 2003: 33–5 F3). When he applied for a travel grant from the German Archaeological Institute in Berlin, he asked Brunn (who was on the board) to disclose what the other board members were saying about him behind his back, notably his old teacher Johannes Overbeck whom he feared disliked him (Hofter, 2003:

42 F8). Later in life his bad temper and paranoid tendencies invariably resulted in highly problematic work situations and strained relations with superiors and colleagues who accused him of being despotic and tactless. Many harsh words were in fact uttered behind his back (Bissing, 1907; Furtwängler, 1965: 229*f.*). His students may have called him the 'Linnaeus' of the discipline, but to some of his colleagues he was more its 'Attila' or 'Napoleon' (Reinach, 1907b: 327).

It is necessary to briefly outline some of the incidents that Furtwängler later claimed had sabotaged his career. Most of them occurred during or as a result of his extremely productive but conflict-ridden years in Berlin, where he was never promoted at the museums, never offered a chair at the university (although he became affiliated as *extraordinarius* in 1884) and never elected to the powerful board of the Archaeological Institute. These three institutions were the major structuring forces in German archaeology at the time. Even if he himself may have suspected anti-Catholic sentiments, which were certainly strong in Prussia at the time, internal museum correspondence indicates that it was rather his own inability to work smoothly with his superiors and co-workers that contributed the most to the frosty atmosphere and strained relations at the Berlin Museums. Furtwängler began this decisive period in his career there as assistant curator to Alexander Conze in the prestigious sculpture department. Conze however took an almost instant dislike to his new assistant, whom he found arrogant and despotic towards his fellow workers, and soon had him transferred to the much less prestigious Antiquarium, directed by the more sympathetic Ernst Curtius (Furtwängler, 1965: 113*f.* nos 60–61; 229*f.* nos 158–61). This was *de facto* a significant step down. Curtius, who already knew Furtwängler well from Olympia, showed great patience with the haughtiness and frequent mood swings of his new assistant and gave him relatively free rein, especially where new acquisitions were concerned (Curtius, 1958: 215*f.*). It is interesting to compare Furtwängler's letters to his mother and sisters (Furtwängler, 1965), which mention none of these conflicts and describe his colleagues as 'fine people', with contemporary work correspondence and later statements by friends and colleagues, where conflicts are clearly spelt out (especially Conze and Bulle in Furtwängler, 1965: 229–32; Hauser, 1908). Struggling on, nevertheless Furtwängler was soon among the most prolific and well-established staff members. Apart from his day-to-day curatorial work, he worked on two ambitious catalogue projects of the museum's 4,000 vases (Furtwängler, 1885) and 12,000 gems (Furtwängler, 1896). He moreover completed his own ground-breaking work on Mycenaean pottery (Furtwängler and Loeschcke, 1886) and bronzes from Olympia (Furtwängler, 1890), and also published a catalogue of the Saburov private collection (Furtwängler,

1883–7). He moreover produced several entries for Wissowa's *Real-Encyklopädie* and a steady stream of minor publications and review articles.

As Raimund Wünsche has noted (2007: 301), even one of his many monographs, several of which were produced during the Berlin years, in itself could well have represented a life *opus* for other archaeologists. Furtwängler was a restless individual – 'everything is fire in him', Brunn is said to have once exclaimed (Curtius, 1958: 214). He hated idleness in himself and in others, including his children, who always had to keep themselves busy (A.E. Furtwängler, 2005: 17). An early riser, he was always the first to arrive at his workplace in Berlin. 'It is very pleasant', he wrote to his mother, 'to be working undisturbed on such great material. Three assistants are at my disposal, *ever ready to obey all my commands* (Furtwängler, 1965: 32*f.* no. 16, my emphasis). Work became even more satisfying when his superior Conze was not around: 'Conze is in Paris and *I reign alone*' (Furtwängler, 1965: 66 no. 31, my emphasis). Always working on several manuscripts simultaneously, writing very fast and seldom revising, he finished at least seven pages on a good day (Curtius, 1958: 214). The manuscript for the Berlin gem catalogue, for example (in the Antikensammlung PKB), is written in a fluent hand with very few later corrections. To this can be added a massive professional correspondence.

When Conze retired from the sculpture department and Carl Robert from the archaeology chair at the university in the late 1880s, Furtwängler saw himself as the obvious candidate for both – or at least one – of these positions, and for good reasons (Curtius, 1958: 215*f.*). He was already extensively published, even if some of his best works still lay ahead of him. He had been responsible for a series of significant acquisitions and catalogues, and had moreover turned down offers from the lesser universities at Erlangen, Rostock and Münster, for which he considered himself over-qualified (Straub, 2007: 43). But instead both positions went to his former habilitation supervisor, Reinhard Kekulé, back then a renowned sculpture expert but for whose work Furtwängler had very little respect. Kekulé was called to Berlin at the express wish of Kaiser Wilhelm II, and it has been remarked, with some justification, that Furtwängler might have been considered not yet sufficiently qualified or suitable for such positions at leading Prussian institutions (Zazoff, 1983: 213*f.*). He nevertheless continued working at the museum for another six embittered years, but seems to have lost no opportunity to cast aspersions on his colleagues, making his own situation there impossible. Contemplating applying for the Yates chair at University College, London, which became vacant in 1888, he asked Brunn for a letter of recommendation (Furtwängler, 1965: 161*f.*

no. 96). Before Brunn had the chance to reply, however, Furtwängler learnt that his old mentor had intended to recommend someone else and he immediately withdrew his request (DAI Zentralarchiv, Nachlaß Brunn). In 1894 Brunn died and Furtwängler was at last called to take up his chair in Munich. But the vilification of his old workplace continued long after Furtwängler had left Berlin for Munich, incidentally with excellent references from both Conze and Kekulé, who no doubt wanted him out of the way (Zazoff, 1983: 214). Things got so bad that the only favourably disposed colleague he had left at the Museum, Curtius, saw himself forced to repeatedly rebuke Furtwängler in the strongest words:

> If you would lend me a friendly ear, then stop speaking of any bad experiences that you have had here! You had *all* that a young scholar could have wished for, in plenty!... Of what can you complain? If your superiors and colleagues were not sympathetic towards you, then who is to blame? (Curtius in Furtwängler, 1965: 195*f.* no. 132)

> If you will accept even the slightest advice from a fatherly friend, then stop all these outbursts of bitterness! We only *poison* our earthly existence with them. Where so many people come together in one place and with similar duties, there is bound to be friction. You are not exactly blameless yourself, since you have irritated some of your colleagues. By and large you have *nothing* to complain about Berlin. What other place could have prepared you better for the Munich chair! (Curtius in Furtwängler, 1965: 201*f.* no. 134)

This drawn-out conflict no doubt contributed to the decision of the Archaeological Institute, headed by Conze, not to elect Furtwängler to Brunn's vacant place as the Bavarian delegate on the board, which would have been the logical choice as Furtwängler was Brunn's successor to the Munich chair. Instead they called a Prussian-born but Munich-based philologist, Wilhelm Christ, who had no archaeological experience whatsoever. A few years later, the board once again sidestepped Furtwängler when they elected a junior and much less qualified delegate from Würzburg to take up Christ's vacant seat. When the call finally came, just a few months before his death in 1907, Furtwängler declined (Hauser, 1908: 469; Kekulé, 1908). There were several powerful personalities on the board who were critical towards and disliked Furtwängler, among them Conze and Kekulé. But resistance in Berlin was more widespread than that. The British archaeologist John Marshall, who visited Berlin in 1894, reported in a letter to his partner Edward Perry Warren that 'they hate Furtwängler very much here' (quoted in Burdett and Goddard, 1941: 187).

And here is where resentment and frustration come to interfere with Furtwängler's scholarship and interactional strategies *vis-à-vis* the scholarly community. The deliberately polemical *Meisterwerke* study of Greek sculpture (Furtwängler, 1893), written in great haste and published in 1893, just before his departure from Berlin, was partly conceived with the aim of discrediting the work of his peers and embarrassing them, especially some colleagues in Berlin, and establishing himself as an undisputed authority in the field. The book reconstructs lost masterpieces of Greek sculpture through careful studies of written sources and especially Roman copies. Although highly influential, the method was and has remained controversial. But it opened up a whole new field of research to scholars: the systematic study of Roman copies (albeit not for their own sake), which had so far received very little attention from academia. In addition to being hailed as a 'Bible for archaeologists' (Hauser, 1908: 468; Sieveking, 1909: 126), the book was also called a '*Kampfbuch*' (an intentionally polemic publication) (Curtius, 1958: 218) and a 'sewer' (Kern in Furtwängler, 1965: 230 n. 162) for its many vicious attacks on the work of his colleagues. Kekulé waited until Furtwängler was safely out of Berlin before countering with a long and sarcastic review (Kekulé, 1895), and full war broke out. This attack was never forgiven and was countered with intense hatred (Curtius, 1958: 218). When a few years later Furtwängler submitted the manuscript of a Berlin catalogue that he had completed in Munich, his former employer requested that he remove or rephrase some unacceptable comments about other Museum employees (Platz-Horster, 2005: 23 n. 51).

Furtwängler's next great project, on ancient gem-engraving (Furtwängler, 1900), was chosen specifically because the many fake ancient gems that had flooded the market for centuries had made gem studies a veritable minefield for serious scholars and collectors. No experts of any note had emerged in the previous century and the task of bringing order to this field called for an exceptionally dedicated individual or group of scholars. If Furtwängler could pull it off single-handed at such high stakes, the project was sure to make his success. And he did. The publication was hailed as one of greatest achievements of German *Wissenschaft* (Willers, 1901: 1103), '[h]istory writing in grand style' (Curtius, 1958: 217) and a book that would remain 'the outstanding work of reference' for the new century (Babelon, 1900: 446). Again, Furtwängler classified a vast material and made it available for further research. Of the more than 50,000 originals that he examined, he published 4,000. There were very few peers competent to review the work. Furtwängler knew this and tried his best to discredit any competitors, positioning himself as a pioneer in the field where he had no modern forerunners of any note, misrepresenting scholarship that he in fact built on more than he was willing to admit (Furtwängler, 1900: III, 402–34). Although in general this way of interacting seems to have been more complex than just simple acts of self-elevation and other-derogation, in the cases where competition could be found or expected, there is little doubt that Furtwängler frequently engaged in deliberately aggressive polemic: his books and review articles often contain excessively harsh judgements, personal attacks or unwarranted invectives. Savage comments like 'unscholarly', 'trivial and worthless', 'a miserable and wholly worthless book of a dilettante' and 'superficial and useless' abound (random examples from Furtwängler, 1900). Such tendencies were already apparent in his earliest publications, for example these characteristic lines from a review article:

> [T]he unfortunately quite consistently unscholarly nature of this new publication cannot be excused, as there is very little here for scholars or for teachers and students: because he who wants to be teacher, must first learn himself (Furtwängler, 1875)

Furtwängler of course was not alone in criticising the work of colleagues; the culture of peer criticism in Germany at the time was severe compared with today's standards. But his tone and choice of words were widely considered to be unbalanced and unjustly brutal, and they were not confined to reviews and publications alone. An American colleague who had briefly studied under Furtwängler in Munich stated that he had a keen sense of the ludicrous, but usually checked himself in the cases when his laughter risked offending a student:

> To the views of others he devoted little time, unless they were acceptable, and when any theories did outrage to his judgment or his artistic sense he passed them by with epithets such as 'completely wrong', or 'inconceivable'. Hence, in part, came the injured feelings of others working in the same field. (Church, 1908: 65)

This was mildly put. It was an extremely risky strategy that could have backfired much more than it actually did. It is perhaps worth quoting the opening lines from Georges Perrot's review of *Die antiken Gemmen* (Furtwängler, 1900):

> In the world of archaeology, Mr Furtwängler is more feared than loved. That is because I know of no one who makes less effort to conceal his scorns that are provoked by the impertinent people who dare to offer their opinions on issues they know little about, and by clever people who with the slightest of skills pretend to be oracles and attain high status. He is at his most cruel when someone has the misfortune not to be on his

> side in matters that strongly engage him. He then crushes his challenger with a contempt that sometimes erupts into violent words, but which more often is to be found there all along in the argumentation and which seems to imply that the adversary should not even dare to think about replicating and prolonging the debate. (1900: 475*f.*)

and Percy Gardner's obituary in the *Classical Review*:

> But there is another side to his work on which I insisted when he was in the field, and which I must not pass over even while we are regretting his loss. In the case of so remarkable a personality, the 'personal equation' was likely to be prominent; and in fact this greatly diminished the value of his work. His antipathies often carried him away; and as a controversialist he seems to have sacrificed love of truth for love of victory. Nor can it be maintained that his judgment and sense of proportion were in any way equal to his knowledge and force. (1907: 252)

Goffman (1959: 163) observed that 'renegades', as he called them, 'often take a moral stand, saying that it is better to be true to the ideals of the role than to the performers who falsely present themselves in it'. Some critics have tried to explain Furtwängler's aggressive way of interacting as resulting from his taking most seriously his role as *Wissenschaftler* and that his intense engagement with the issues that really interested him made him somehow incapable of realising fully that his harsh criticisms could offend (Bulle in Furtwängler, 1965: 23*f.*; Curtius, 1958: 222). Others are more inclined to view these attacks as deliberate strategies, notably Gardner (above), who in my view is right in assuming that they had negative effects on the reception of his work. While Furtwängler in all likelihood did not act aggressively simply to create dissonance and awkwardness, he may well have done so in the knowledge that conflict would be a likely result, thus destroying or threatening the polite appearance of harmony and consensus in the field (Goffman, 1959: 205). Either way, the kind of unbalanced polemic that Furtwängler engaged in closed many professional doors and valuable information channels. To compensate, he repeatedly stressed the primacy of original artefacts and the need to always start afresh and from scratch, trusting only one's own eyes. This standpoint allowed him to disregard or dismiss much of the work of his colleagues, on which he in fact drew considerably more than he chose to acknowledge (e.g. Willers, 1901: 1172; Zazoff, 1983: 228). This could give the false impression that Furtwängler had a more direct or unbiased relation to the objects he studied. But although he undeniably had a very good command, often unrivalled first-hand knowledge, of the material categories that he studied, his gaze was of course coloured by a host of preconceived standpoints (see especially Bažant, 1993).

Although he had dedicated his *Meisterwerke* (Furtwängler, 1893) volume to his mentors Brunn and Curtius, when he was later asked what he had learnt from his teachers, his answer was predictably 'nothing' (Reinach, 1928: 204). Nevertheless, it is clear from the reading of any of Furtwängler's many publications that he had the greatest respect for the discipline and its traditions, and that he was keenly aware that what had been achieved by his forerunners and contemporaries was what he continued to build on.

Having, willingly or not, turned against or offended many of his colleagues who worked on the same material, Furtwängler refocused his attention and network building on people who handled artworks: museum curators, auction houses, agents, dealers and private collectors, who all figure prominently among his correspondents. While never actually leaving any of the archaeological networks that he belonged to – that would have been unthinkable – this wider group of contacts now became his real powerbase, because they could provide him with highly desirable and hard-to-get information. He had had excellent opportunity to collect and cultivate them during his Berlin years, when he was expected to attend international auctions and sales and negotiate new acquisitions with dealers and collectors on behalf of his employer. He had also visited most of the major museum and private collections in search of *comparanda* for his own catalogues of sculpture, vases and gems (e.g. Curtius, 1958: 215). Furtwängler thus met and befriended more or less everyone worth knowing in the art market and museum world and became fairly fluent in French, English, Italian and modern Greek (Wünsche, 2007: 302). By freely offering expert opinions in return for hard-to-get documentation and information about unpublished artworks, he made sure that no new items of any note were excavated, changed hands or came on the market without his knowledge. During his Munich years, in matters of authenticity and attribution Furtwängler's opinion was regularly sought and seldom disregarded (e.g. Moltesen, 2012: 166*f*., 197, 199; Rouet, 2001: 36–40; Dyson, 2004: 102). We are told that when Furtwängler left Berlin for Munich, he brought with him all these useful art market contacts and turned Munich into a major caravan station for the international antiquities trade (Schuchhardt, 1956: 17). His one-time assistant and on-and-off associate in Munich, Paul Arndt, was both collector and art agent for notable museums and collectors like Carl Jacobsen (Moltesen, 2012; forthcoming), and Furtwängler's transfer to Munich meant a great boost, not only for the university but for the museum collections that came under his direction, as well as for the local community of collectors. Leading museums, dealers and private collectors sent whole sculptures, vases and gems to be assessed on a costly detour to Munich – if not the originals themselves, then casts or very good photographs, documentation which Furtwängler kept for his private archive. Large investments could depend on his judgements, careers be made or broken. He certainly had an excellent visual memory and a critical eye on which to base his opinion, but of course he was not infallible and again one might suspect that his strong antipathies sometimes interfered with his judgement. Possessing information often unavailable to his rivals, Furtwängler made sure his own research would not be too dependent on them, but that instead they would have to turn to him for information from his vast personal archive and extensive intelligence machinery. The many letters that Furtwängler received from fellow scholars, today in the care of the German Archaeological Institute in Berlin, are full of questions and requests but very few answers to specific queries, which suggests that Furtwängler himself relied on alternative information channels. So it seems we do not have the ideal flow of information here between generous colleagues, but rather brief notes on a need-to-know basis. More than 80 per cent of Furtwängler's great mass of correspondents exchanged fewer than five missives with him, often brief notes or postcards. Only a handful of people were regular correspondents, but their letters were never numerous.

While doing fieldwork on Aegina, Furtwängler contracted dysentery and was transferred to Athens where he died a few days later, on 10 October 1907. His funeral was attended by several ministers and dignitaries and a speech was made by the director of the French School in Athens, Maurice Holleaux, in the absence of Dörpfeld who still directed the German Institute. Obituaries and eulogies were published in Munich newspapers and journals (notably Bissing, 1907; Bulle, 1907; Studniczka, 1907; Sieveking, 1907; Hauser, 1908; Wolters, 1910) and around Europe (e.g. Reinach, 1907a, 1907b; Gardner, 1907; Mach, 1907; Church, 1908). But the Berlin institutions, newspapers and journals were for the most part silent; the *Archäologischer Anzeiger* published a short official statement by Kekulé (1908) on behalf of the Institute. Heinrich Bulle, who had studied and worked with Furtwängler, confessed in a private letter that it was not until after Furtwängler's death that he realised that he had in fact been very fond of him. 'After each conflict we had... I felt drawn even closer to him' (Bulle in Furtwängler, 1965: 231*f.*). A few years later, a short but significant book called *Ägineten und Archäologen: Eine Kritik* was published by the Strasburg archaeologist Maximilian von Groote (1912), which is nothing short of a detailed but unbalanced critique of, not to say attack on, Furtwängler's whole contribution to archaeology. Groote concludes with the remarkable statement that no contemporary had done more damage to the study of Greek art than Furtwängler, and that, human sentiments apart, his death was no loss for the general progress of *Kultur* (1912: 88).

The example outlined here is in many way ways extreme, but conflict and friction are constant components of knowledge production and its contexts, affecting where, why, how and by whom knowledge is produced, disseminated, accepted or rejected. It is the task of historians of archaeology to trace and critically assess these and other time-, space- and context-specific factors that potentially affect the dynamic processes of the discipline. Furtwängler operated in, and at times against, structures, networks and clusters that he perceived as partly hostile but to which his scholarly production nevertheless constituted a direct response and his part in an ongoing conversation that he fully recognised and sincerely believed in. He also found alternative networks and clusters where he deemed the extant ones inadequate or unsympathetic or uncooperative. Where face-to-face interaction was difficult or deemed impossible, his own published works became his chief instrument of interaction. They contain much of the sort of criticism and friction that are otherwise mostly ventilated more informally at congresses and conferences, which Furtwängler of course rarely attended. These published works and the powerbase as oracle that he built for himself guaranteed a key position in this 'field of struggle', despite resistance from some formidable enemies that he had made on the way. The stakes had been considerable; Furtwängler risked not only being socially isolated, which he certainly was in some circles, but becoming intellectually marginalised. A less creative and productive scholar might not have been as successful. Instead, Furtwängler's work came to have a substantial impact not only on his contemporaries but on the next two generations of German Classical archaeologists, and also outside Germany. In addition to his own work, part of his lasting success can be ascribed to his small but close-knit group of devoted students and followers, some of whom came to be highly influential during the first half of the twentieth century: Paul Arndt, Ludwig Curtius, Hermann Thiersch, Friedrich Hauser, Johannes Sieveking, Heinrich Bulle, Walter Amelung, to name just a few. Brunn had made Munich an important centre for the study of especially ancient sculpture; his student Furtwängler transformed it to the very heart of so-called *Stilarchäologie* and *Kopienkritik*, building an outstanding archaeological library, cast and photograph collections, and challenging Berlin's leading position in Classical archaeology. It would perhaps not be an exaggeration to identify the 'Munich School' of *Kunstarchäologie* and *Kopienkritik* which Furtwängler created as one of the more influential and enduring 'Thought-Collectives' (Fleck, 1979) of the discipline. I end with Percy Gardner's words from his obituary of Furtwängler in the *Times* (15 October 1907):

> When he was in the field of controversy, no one else seemed to be worth noticing. His preponderant force held the field, and seems to have

reduced almost to silence the majority of his German colleagues, who will be disposed to write on his tomb the line of Pope: *The great, the fierce Achilles fights no more!*

Notes

1 18 August 1878. DAI Archive, Nachlass Brunn. *Cf.* also Furtwängler, 1965: 174*f.* n. 105*f.*
2 DAI Archive, Nachlass Brunn; Hofter, 2003: esp. 29 F1, 43 F9.

8

When the modern was too new: the permeable clusters of Hanna Rydh

Elisabeth Arwill-Nordbladh

The production of knowledge – some theoretical considerations

The production of knowledge is a social process, linked to various premises. Exploring the emergence of scientific knowledge involves not only investigations of the research community and its diverse conditions but also studies of the individual researcher – a biographical perspective. Bonds of partnership, loyalties and shared scientific ideas, or their opposite, distrust and ideas called in question, can be mapped out as more or less formalised and stable webs or networks. As agents, the individuals or groups in such a network may operate in inter- or intra-relational affirmative negotiations. Such groupings have been studied by, among others, Bruno Latour, using the framework of actor-network theory (ANT) (e.g. Latour, 2005). Following the actor-network perspective of Latour, not just people are involved in such negotiating processes, but also material phenomena such as physical objects, and immaterial factors like virtual realities. Just as do the individual agents, these non-human features, which Latour calls actants, constitute an active power in the dynamics of negotiation.

For many archaeologists, the notion of a mutual, active, agential force between human agents and non-human (including animal) physical phenomena is not difficult to endorse, as varieties of this perspective have accompanied archaeological thinking for several decades (Gillberg and Jensen, 2007: 11), for example via the post-processual ideas regarding the mutual relationship between subject and object as formulated by Ian Hodder (e.g. Hodder, 1986), the notion of Arjun Appadurai's (1986) social life of things and Alfred Gell's secondary agency (1998), and more recently the agential realism of Karen Barad (2003) and the material symmetry of Bjørnar Olsen (2010). Even if there are considerable theoretical differences between these perspectives, they share the idea that human agency and material phenomena's agential capacity connect in inter- and intra-relational dynamics in various ways, on various agential levels and with various force. However, one of the distinctive features of ANT is that such analyses are based on the assumption that the agential dynamics constitute systems, possible to study on various social scales, in which the agential subjects – the human agents and the material actants – are parts of one or several networks. Professional networking is crucial to the production of knowledge.

One fundamental element in knowledge-producing processes is spatial location; the geography or landscape of knowledge (Livingstone, 2003, 2010). The geographical approach can be understood in a literal sense, such as the spatial distribution of clusters and nodes where science is performed. This includes communication facilities and other factors that encourage networking. It can also be seen more figuratively, for example as the knowledge-producing room (Livingstone, 2010). From a historical perspective such explorative arenas can be identified as informal spaces such as market places, workshops and even kitchens. During the professionalisation process of various disciplines, like antiquarianism's path towards the discipline of archaeology, more formalised arenas emerged, such as museums, laboratories and field sites, all with their particular professional codes. They were all shaping situated landscapes of knowledge production through the dynamics of networking.

Against this background several questions may be posed. Who had access to the arenas of knowledge production? Were there professional borders that were open or closed to certain individuals and groups, for example based on wealth, gender or colour of the skin? In what way did such different circumstances affect a discipline in terms of research questions, economic support and the acceptance of scholarly results? Questions like these have been much debated within feminist scholarship. Many researchers (for example Harding, 1991; Haraway, 1992; from an archaeological perspective Conkey and Gero, 1997) conclude that scholarly research is situated. Individuals and groups are positioned at specific points on different axes of power, such as gender orders, economic structures, various bodily abilities, racial, ethnic or religious identities. For the individual subject these positions are intersections, possessing a dynamic status from a power perspective. Such situated positions also affect networking processes and thus knowledge production.

A biographical approach

In this chapter I will discuss a particular scientific contribution made by the archaeologist Hanna Rydh (1891–1964). As the first woman in Sweden to achieve an archaeological doctorate, she had to navigate within a male-oriented discipline, which was developing its professional identity. Striving to earn her place in the Swedish archaeological community, Hanna Rydh's professionalisation strategies were situated in various circles. In addition to the archaeological scientific cluster, Hanna also was affiliated to feminist groups striving for female emancipation both within the academy, and in society as a whole. Being an academic in the early-twentieth century, Hanna made the quite unusual life choice to marry a colleague and raise a family with children, while maintaining her archaeological activities. Throughout her academic life, her family constituted a supportive circle.

The theme of the present discussion is Hanna's encounter with some social and professional networks of the 1920s, namely a national and transnational circle promoting women's emancipation, a specific research milieu in the French archaeological national museum at Saint-Germaine-en-Laye near Paris, and the scholarly cluster of the Museum of Far Eastern Antiquities in Stockholm. The main result of her stay in Paris was a book about Palaeolithic cave art, favourably received by Swedish readers (Rydh, 1926a). A few years later she wrote two articles about ceramics and fertility rites that were dealt with somewhat cursorily by the Swedish archaeological research society (Rydh, 1929a, 1931). I here propose that Rydh, through her contacts with a specific French archaeological milieu, gained inspiration for these texts from up-to-date scientific theories of social arrangements, rooted in the sociological school of Émile Durkheim. Unfortunately, the contemporary Swedish scholarly world was unable to see the advantages of her approach.

The empirical evidence is looked at from a biographical perspective. Focus is on the biographic subject as an active agent. With this perspective, the tension between agency and structure is obvious (Berghahn and Lässig, 2008), including networking, networking's geography and knowledge-producing rooms.

Hanna Rydh, a short presentation

Born into a wealthy family, Hanna's childhood and youth seem to have allowed her to develop her natural gifts, which included an endowment for studying languages, cultural interests and sports. Her father was a successful engineer, managing a prosperous family business. Her mother had been a teacher before marriage, a fact that certainly indicates a positive attitude towards women's education. The family, which included Hanna's older brother and sister, who was also a trained teacher, formed an intellectually stimulating and encouraging setting. Hanna's school records demonstrate that she was an interested and talented pupil. As early as her mid-teens, she was active in the local branch of a national youth association. There she gained a solid experience of club activities, such as suggesting issues for the agenda, and proposing motions from the rostrum before large audiences. These must have been useful lessons for her future work in the academic, cultural, social and political worlds (Arwill-Nordbladh, 1995, 1998, 2005a, 2005b).

The emancipation of women

A frequently debated question of those days, in the vocabulary of the time, was the women's issue. In Sweden, until 1922 a married woman was placed under her husband's guardianship. This meant that his was the privilege to determine his wife's economic business, speak for her in legal affairs and decide upon family matters like the way of living, children's upbringing and other issues of importance. Another urgent question was women's access to education. Having completed primary school, most girls and boys attended separate schools, following different curricula. This often meant that, after finishing school, students of different genders had different qualifications, and for girls many doors to future work or education were closed. However, a few schools or tutorial systems followed a curriculum that provided the qualifications necessary for university entry, and Hanna's school was among these. From 1873 women were allowed to attend some university disciplines, but nevertheless progress was very slow and not until ten years later did the first woman obtain a doctorate. When Hanna received her doctoral diploma, only twenty-three women in all had reached this goal in Sweden (Markusson Winkvist, 2003: 232). Still, the most important issue for the women's movement was the franchise. The demand for women's right to vote had been on the agenda since the last decades of the nineteenth century. In 1909 all Swedish men were given the right to vote, but not until a decade later was the same civil right extended to women, and in 1921 women could submit their votes for the first time.

Looking at Hanna Rydh against this background, she can be seen as a good representative of the pioneer group of young women who, with their own lives as example, strove towards women's emancipation in education and academic work, and their participation in social and political life. Informal groups in the academy and the national suffragette movement that constituted Hanna Rydh's female networks were of great significance to her, and in a mutual interaction she also had an influence on them.

The significant clusters of Hanna Rydh's formative years

Of the documents that tell us of Hanna Rydh's life, her pocket diaries are of particular relevance (Gothenburg University Library, KvinnSam, National Resource Library for Gender Studies, Hanna Rydh Archive A 12). Here the calendars from the 1910s and 1920s record important events, sometimes on a day-to-day basis, showing her strategies and negotiations to gain a place in the discipline. The professional sphere of contacts ranged from a small group of antiquarians to a more organised arena with local, regional and national institutions. Discussions and debates in periodicals and other publications were frequent, and peers and colleagues guaranteed the quality. Hanna met and took part in an archaeology which was proceeding to find its shape as an academic profession (Arwill-Nordbladh, 2005a: 114–15).

In 1910 Hanna Rydh graduated from school, and soon she registered at Stockholms Högskola. This university college, later Stockholm University, was at that time a private educational institution, renowned for its modern ideas – reflected for example in the appointment of the Russian mathematician Sonia Kovalevsky to a professorship as early as 1884. Hanna signed up for studies in the Humanities, within the subjects of literature, history of art and archaeology, focusing on the latter. In 1914 she passed her *Laudator* in archaeology, and after finishing her studies in medieval art a year later she was one of the 435 Swedish women who had obtained their bachelor's degree. She was now ready to take a place in the archaeological world. Soon she was linked to The National Heritage Board for temporary excavation projects such as a research project connected to the emergence of the Kingdom of Sweden. The research questions were of significance for the national historical narrative, and at various intervals the project went on for almost a decade. Being given charge of the prehistoric section is a clear indication that Hanna was a respected and trusted colleague within the research community.

After graduating, Hanna registered as a senior member at the University College, starting her PhD project. As a member of the first generations of female university students, Hanna and her female co-students performed what can be labelled a double border crossing (Markusson Winkvist, 2003: 33). By diverging from the traditional female gender role and at the same time challenging the masculine academic role, they had to create their own identities and practices. Their life was a world of negotiations of gender positions. This, in the Swedish context, often meant having to choose between adapting to the norm of the mainstream professional, or separating from the general model by embracing so-called feminine values. During Hanna's formative years of academic schooling we can see that she took both positions.

As archaeology was a rising discipline, its members were developing various ways to promote the scientific character and quality of the subject, in order to professionalise the field. For Hanna, as a young student, it was important to follow the academic route in the proper way. The pocket diaries tell us that attending lectures, examinations, excursions and minor field training expeditions were mixed with days of intense study at the library and the museum. This gave Hanna opportunities to create friendship-based networks with her fellow students, something that turned out to be useful in the years to come. Some of these groups, such as The Society for Students in the Humanities, The Association of the Students of Stockholm, the Society for Art History and the Archaeological Club, were gender-mixed. Within these circles, Hanna took an active part in discussion evenings and other social events, thus getting the opportunity to get familiar with academic customs. Nevertheless, she must have felt the need for a place which could highlight the conditions of female students, as after spending one year at the university college, she helped found The Women's Student Association in Stockholm. One of its aims was to serve as a club for 'discussion and support' in a friendly atmosphere (Hallind, 2004: 203). This female support was reinforced when Hanna, after receiving her bachelor's degree, joined the ABKF, an association to promote women with a university education (Fridh-Haneson and Haglund, 2004).

This dual-strategy position can be seen in Hanna's written work. In her explicit archaeological endeavour she, in tandem with her colleagues, was engaged in projects towards a modern, scientific and scholarly discipline. However, in her popular texts (e.g. Rydh, 1926b) she developed research that differed from mainstream scholarship and was more in line with her feminist interests – as early as in 1927 she used the word 'feminist' to characterise herself (Rydh, 1927: 11).

In Stockholm during this period it was not too difficult to get in contact with central figures regarding liberal progressive ideas. At the World Peace Conference in 1910, Hanna's pocket diary tells us that she listened to Oscar Montelius who gave the introductory keynote speech, and to the social reformist and author Ellen Key's speech about women and peace. That Montelius supported these issues is not surprising. He was well known for promoting the emancipation of women (Arwill-Nordbladh, 1987; and see chapter 6). For almost two decades his wife Agda was the president of Fredrika Bremerförbundet, in those days Sweden's foremost women's association. When Hanna had reached the stage of finishing her dissertation, Montelius gave her vital support in the process of getting it printed in time (Arwill-Nordbladh, 1995: 80–81).

The active position of the women's issue is shown by the fact that the Landsföreningen för Kvinnans Politiska Rösträtt (the Swedish assembly for votes for women, LKPR) hosted the congress for international women's suffrage in 1911. Hanna participated as one of the student stewards in the festival procession on the final evening. In this way her female students' network was linked to the top international suffrage groups. Another example of how the networking campaigns for the emancipation of women connected the local with the national can be inferred from a note in Hanna's diary from April 1913, when she wrote: 'LKPR, name raising petition, Dj'. This suggests that the nationwide project organised by the LKPR to collect the names of supporters met in Djursholm, the garden city and Stockholm suburb where Hanna lived. The campaign resulted in a petition to the Swedish Parliament with more than 350,000 signatories (see Figure 8.1).

During her studies, Hanna found a kindred spirit in Bror Schittger (1884–1924), antiquarian and associate professor in archaeology at Stockholms Högskola. In 1919, after successfully gaining her doctorate, Hanna and Bror got married. Their collegial network had now turned to a family-based bond of loyalty. Here the professional and private spheres merged into one single entity, which in an encouraging and supportive way affected the production of knowledge for both of them.

Hanna's professional milieu, constituted by her archaeological colleagues such as the senior Montelius, her academic peers of the same age and the institutions employing them, was for the most part encouraging and inclusive. Maybe because of this, Hanna kept to the conventional research track. However, the women at the University College clearly reacted to the generally unequal social conditions between women and men by forming women-only groups for empowerment and support. In these circles, Hanna was encouraged to write her first popular texts about prehistoric women.

The French experience

This gender-specific support was soon affecting the academic life of Hanna Rydh. In 1922 AKBF joined the International Federation of University Women, IFUW, an association supporting peace and women's access to higher education. The same year the IFUW funded an international scholarship to promote women's studies, for which the members of the Swedish branch could apply. In a letter to Schnittger, Hanna asked

8.1 This emblematic photo from 1913 shows the author Elin Wägner in front of the collection of names for the LKPR's petition for women's votes, handed over to the Swedish Parliament. Hanna Rydh's name is most probably one of the 351,454 signatures on the petition. Photo: KvinnSam, Gothenburg University Library. Copyright © Gothenburg University Library. All rights reserved and permission to use the figure must be obtained from the copyright holder.

> Could you please write and tell me which institutions in France I should list, for studying the Stone Age? Last Wednesday I met Miss Sturtzenbecker [secretary of the ABKF], who asked me to apply [for the international scholarship]. There is no doubt that the Swedish committee will give me their recommendation, as they consider me to be the candidate with the best credentials. (Hanna Rydh to Bror Schnittgers, Riksantikvarieämbetet, Stockholm, Antikvarisk-topografiska arkivet [ATA] vol. 4).

Hanna was awarded the scholarship, and it is significant that she was eager to reciprocate the support she had enjoyed. This can be inferred from two letters sent to Mrs Stina Rodenstam, National Instructor of Domestic arts and crafts. In these Hanna wrote:

> The Association of Academically Educated Women is trying to collect products of Swedish art and Swedish handicraft to send as a gift to a grand bazaar, which will take place late this autumn. One purpose is to establish an international scholarship for studies, granted by The International Federation of University Women – such a scholarship, entirely collected in England, has for the first time been awarded this year, and to my own surprise I received it for studies in France – and the other purpose is to establish a club house in London for female students of various nations...

The call for funding support was sent to 'all countries', and Hanna and her friends were eager to compare favourably with the other fundraisers, so they approached the national network for Domestic handicraft for contributions 'which will be received with greatest gratitude, however small they may be'. A few weeks later Hanna wrote:

> Please accept my warmest thanks for the magnificent delivery! It was more than kind, and we are so grateful. All the objects were so beautiful and we really appreciate the chance to get our Swedish handicraft so comprehensively illustrated. Everyone has been so kind, and the fundraising drive has run so much better than we dared to hope at the beginning.
>
> Our delivery, which will be shipped on Saturday, has a value of more than 2,000 crowns and in addition we can send 1,000 crowns in cash, so now we consider that we don't have to feel ashamed. (Letters from Hanna Rydh to Stina Rodenstam, 2 October and 19 October 1922, my translation).

French archaeology had a good international reputation, notably in the fields of Palaeolithic stone technology and Palaeolithic cave art, and archaeologists from many parts of Europe made study trips to France to acquire up-to-date knowledge of the most recent approaches. One central institution was the Musée des Antiquités nationales at Saint-Germaine-en-Laye near Paris. The curator for the archaeological department was Henri Hubert (1872–1927), an archaeologist specialising in Asian religion as well as Celtic and Germanic prehistory and early history. Hubert also gave lectures in primitive European religion at the École Pratique des Hautes Études and in national archaeology at the École du Louvre. In the 1890s Hubert met Marcel Mauss, who introduced him to the ideas of Émile Durkheim, Mauss' uncle. Hubert and Mauss became close friends and joined in creative collaborations. They were both deeply engaged in Durkheim's project the periodical *L'Année Sociologique* (Mauss, 1983: 149; Isambert, 1983: 154; Schnapp, 1996: 59–60; Schlanger, 2006). Within the group around Durkheim, Mauss and Hubert were assigned the task of directing and developing Durkheim's sociological perspective on ethnography, the ethnography of religion, history, archaeology and prehistoric religion (see e.g. Besnard, 1983: 27). In these matters, it was important to chisel out the characteristics of myth as a social element.

From Hubert's Durkheimian perspective, religion, the sacred and, in particular, myth were in focus. Myth was understood in a broad sense. This included its attachment to religion, folk belief, collective representations – something that connected myth to a particular social context – all possessing specific functions within society and myth's creative capacity (Strenski, 1985: 360–61). One of the most important statements was that myth operates through ritual behaviour; the ritual articulated the myth. Strenski quotes Hubert's 'brilliant metaphor' from 1919: 'Myths are social products; it is in the rituals that society is visible, present or necessarily involved. The mythological imagination dances on the threshing floor trodden by rituals, and it is here that one might grasp it' (Strenski, 1985: 362).

According to Hubert's understanding, one fundamental aspect of the relation between myths and rituals was the latter's recurrent reiteration. Such repetitions demand a specific understanding of time, and Hubert developed a theory of mythical time, a concept that differed from the ordinary, mundane time (Isambert, 1983: 157–60; Strenski, 1985: 365–6). Between the rite and the myth is a temporal connection, and this relation presupposes a particular representation of time, a representation which creates a temporal milieu. This milieu, with its particular characteristics, structures the rite. Within the temporal milieu occur specific, crucial days that disrupt the continuity of time, like for example days of periodic, often seasonal festivity, thus 'entirely contaminating' the conventional time with its particular qualities (Isambert, 1983: 158). Among other things, Hubert propounded broad perspectives for an understanding of seasonal festivities, something we will see was of importance for Hanna Rydh's articles (1929a, 1931).

The scholarship gave Hanna funding for six months' studies in France, with the formal position of attachée at the museum and Hubert as her supervisor. The decision about the scholarship was made in July. A few months earlier, Hanna's and Bror's oldest son was born. This caused some worries for the London-based scholarship committee, the British Federation of University Women, who sent a telegram asking if she intended to keep her grant. 'The committee clearly felt torn between anxiety over the child and concern for Mrs Rydh's research' (Arwill-Nordbladh, 2005a: 133).[1] Hanna's answer came promptly: '[m]y son's birth makes no difference'. 'This laconic wording caused great mirth in British circles, and became more or less an anecdote' (Bang, 1931, unnumbered). Hanna left for France in the autumn that same year, organising for Bror, the baby and a nanny to join her soon after. At the museum, Hubert signed Hanna's *carte de travail* which for six months gave her free admission to all the museum's exhibition rooms during the week's *jours d'étude*. She also received entry passes to other museums and Le Bibliothèque National and planned to follow lectures in archaeology and ethnography at the Sorbonne. We can see how Hanna's scholarly 'room' expanded considerably. However, these plans had to be altered. A few weeks after Hanna's arrival, she was urged to return home because of the illness, and soon after death, of her mother. In spring the following year Hanna returned to Paris for a few months. She was now expecting their second child, and in the summer another son was born. By this time it was obvious that Schnittger's health was frail, and eventually it became clear that he had an incurable disease. In June 1924 he passed away. At the age of 33 Hanna became a widow with two small children, 1 and 2 years old.

However, the work at the museum was not yet finished and some of the grant was still available. Hanna wanted to return, and in late June Hubert wrote to tell her that she was welcome back to finish her work. In Stockholm Hanna's father, Schnittger's two sisters and a nanny took care of the children. Before Hanna returned to Paris, Hubert informed her that he had prepared her material so it would be easy for her to recommence her work. In accordance with the French protocol there was also a sanction from the highest level: 'M. Reinach has no objections to your admission… Your key and your equipment wait for you at the office' (Letter from Henri Hubert to Hanna Rydh, 6 October 1924).[2] Quite obviously, Hanna encountered an attitude of warm welcome and she stayed for one-and-a-half months to finish her work. Through these French sojourns, even if they had occurred with intervals, she must have had the opportunity to follow Hubert's thinking for two years.

Returning home, Hanna used her experiences in French Palaeolithic scholarship to write a book about the archaeology of the Upper Palaeolithic, attributed to the genre of popular science. However, it was also read in professional circles, and for many decades it was the only book in Swedish about Palaeolithic cave art written by a professional scholar (Rydh, 1926a).

The East Asian connections

In the years following Schnittger's death, Hanna Rydh spent much of her time finishing some of his archaeological assignments (Rydh and Schnittger, 1927, 1928). She also developed her own archaeological projects, for example guided tours to heritage sites and accompanying books, thus opening out the sites as a gender-inclusive public space (Arwill-Nordbladh, 2005b). She also wrote a book on women in prehistory, travel books and childern's tales (Rydh, 1926b, 1927, 1928, 1929b, 1930). All of these included historical and archaeological themes. Consequently, the 1920s was a productive time for Hanna, demonstrating her ability to master a variety of genres. Unexpectedly, in 1929 and 1931 two works appeared that differ from her previous and later production. In the *Bulletin of the Museum of Far Eastern Antiquities*, a recently established journal from the new museum in Stockholm of that name, she discussed questions concerning the symbolic meaning of ornament design in Chinese and Scandinavian Neolithic pottery (Rydh, 1929a). The results led to further investigations of the mythical meanings of seasonal rituals in China and Scandinavia (Rydh, 1931). These texts are seldom referred to by Swedish archaeologists, and they do not seem to have left much impact on later research. Here I would like to propose the idea that in these writings Hanna made an initial attempt to apply thoughts and ideas that she had encountered in her contacts with Henri Hubert. Thus she tried to transfer some of the Durkheimian ethnological-sociological archaeology that Hubert and his colleagues had developed to the specific research milieu of the Museum of Far Eastern Antiquities in Stockholm. Even though archaeology and ancient Oriental art was the main focus of this museum, scholarships in subjects as diverse as palaeontology and linguistics were also components of importance for its intellectual thinking.

The founder and director of the Museum (in Swedish: Östasiatiska museet) in Stockholm, Johan Gunnar Andersson (1874–1960), was a geologist, palaeontologist and archaeologist (Andersson, 1929a). As a young man he joined two expeditions to the Arctic – basing his doctoral thesis on those experiences – and one expedition to the Antarctic. Appointed director of the Geological Survey of Sweden, he focused on surveying and mapping iron ores. This knowledge led to his invitation by the Chinese government to survey and investigate Chinese mineral resources, in particular to locate iron ore and coal. During his extensive travels, he noticed many locations of mammal fossils and archaeological artefacts. Some of the fossilised bones were collected, and later analysed by palaeontologists. Fossils from the site Chou K'ou Tien were identified as a specific species of *homo*, given the name *Sinanthropus*. By then, Andersson's interest had approached archaeology, with a focus on surveying and excavating. Particular attention was drawn to Neolithic painted pottery ware in the Gansu area in north-west China, bordering Inner Mongolia. In 1925, before Andersson's return to Sweden, an agreement was made to bring the archaeological material, mostly ceramics, to Sweden for investigation. Half of the material would be returned to the Geological Survey of China after finishing this scientific study and description (Andersson, 1929a: 23).

Back in Sweden, Andersson had the ambition to establish 'a research institute devoted principally to the study of prehistoric material from China' (Andersson, 1929a: 26). After successful fundraising campaigns, he reached his goal, backed by a supportive organisation, the China Research Committee, for which the Crown Prince was the chief patron. However, the core of the Museum was archaeological specimens and Oriental art objects rather than fossils. This gave Andersson a reason for establishing the museum's *Bulletin* with the ambition of maintaining a scientific quality comparable to esteemed periodicals like e.g. the *Palaeontologica Sinica* (Andersson, 1929a: 11). While investigating the painted design of the pottery, Andersson found it 'evident that many of the painted designs were magic symbols' (Andersson, 1929a: 26), linked to folk religion and similar perspectives. The range of these approaches urged Andersson to seek collaboration with other researchers, among them Hanna Rydh (Andersson, 1929a: 27).

The point of departure for Hanna's first article, 'On Symbolism in Mortuary Ceramics' (Rydh, 1929a), was Andersson's own observation that the particular ornamentation of Neolithic black-painted Gansu pottery was always connected with dwelling sites, whereas the ornaments painted in black and red, often serrated and with triangular form, were connected with mortuary sites (Andersson, 1929b: 66; Rydh, 1929a: 71). According to Hanna, these patterns showed strong similarities to incised ornaments on Scandinavian Neolithic pottery. To develop these comparative observations, as Andersson expressed it Hanna searched 'very extensively in European archaeological literature and accumulated a vast store of facts bearing upon the problem here in question' (Andersson, 1929b: 69). Hanna's conclusion was that this motif certainly represented a 'death pattern' but simultaneously had significant connections to fertility and life's regeneration, with the ambition to help 'the deceased to a new life' (Rydh, 1931: 69). She thus extended Andersson's proposal geographically, and added the extra meaning of fertility and reincarnation to the symbolic significance. It can be noted that Andersson likewise hints at far-reaching geographic connections (Andersson, 1929b: 66), thus confirming the predominant paradigm of geographically widespread cultural-historic features. This view may also have been the basis for Hanna's study. Today, most likely, a more contextual critical understanding would have been required.

What seems clear is that Hanna and Andersson had discussed the issue when they were studying the material together. Furthermore, it seems that Andersson had agreed on Hanna's comparative observations (Rydh, 1929a: 71), and that he had no objections regarding her conclusions about the connection between death and fertility. Andersson's only demur to the conclusions was that Hanna did not pay attention to his own idea that the serrated design probably had a prophylactic meaning as well (Andersson, 1929b: 66). As we shall see, Hanna returned to this (in her opinion) not particularly significant dispute in her next article. Otherwise, the scholarly consensus was clear.

It is evident that Hanna's article had its background in Andersson's invitation to participate in the *Bulletin*, but that might not have been the only factor. In a letter to Andersson signed 18 May 1929, Hanna's father, J.A. Rydh, offered a donation of 15,000 Swedish crowns to the China Research Committee, to be paid in instalments of 5,000 crowns in 1929, 1930 and 1931 respectively. '10,000 crowns of this sum are intended for purchases for the Museum's collections. The remaining 5,000 crowns is intended to support the archaeological research which my daughter Fil. Dr Hanna Rydh and professor J.G. Andersson plan together' (letter from J.A. Rydh to J.G. Andersson, 18 May [1929]). This letter can be interpreted in different ways. It may be the result of J.A. Rydh's genuine interest in East Asian prehistory and care for the Museum's collections. But it can also be understood as a father's concern about his daughter and her opportunities to develop her scholarly mission. In 1929, 15,000 Swedish crowns was a considerable sum of money, equivalent to more than 400,000 SEK today. Whatever the motivation, it seems to have been a win-win situation for all three of them.

However, this harmonious picture seems to have cracked. In 1931, a letter dated 10 March was written by Andersson to Professor Bernhard Karlgren regarding Hanna's second contribution to the *Bulletin*. Karlgren (1889–1978) was a well-known sinologist, Professor in East Asian language and culture, Vice-Chancellor of the University College of Gothenburg and one of the members of the Museum's Research Committee. The letter runs:

> Dear Brother Karlgren,
>
> Now I must bother you with Hanna Rydh's manuscript, which some time ago was sent back to her, informing her that she must do something to the East Asian part of the matter, in case it should be published in our *Bulletin*.
>
> As you will see, on pages 33–50 she has made a rather passable journalistic combination of de Groot and little Granet, and it seems to me that the best compromise for us would be to swallow the pill, on the condition that we may change the title to something like 'Seasonal fertility rites and the death cult in Scandinavian *[sic]* and China'.
>
> As you know, it is difficult to say no completely, as she among other things has provided us a donation of 15,000 crowns. So, if you please, could you look at it, in particular the new pages, and make whatever remarks you consider are needed?
>
> Yours gratefully (letter from J.G. Andersson to B. Karlgren, 10 March 1931, my translation).

The letter states that Hanna's article was problematic from Andersson's perspective. The text should be seen as a journalistic piece of work, which did not come up to Andersson's scientific ambition for his *Bulletin*, which he was trying to position as comparable to one of the most prominent journals in the field. We understand that the manuscript had been sent back for corrections and that Hanna had rewritten some pages. It seems that Andersson was in a quandary over whether to reject the proposed article.

Fertility rites and death cult – two sides of the same coin

Compared with Hanna's preceding contribution in the *Bulletin*, which runs to some fifty pages and has more than ten plates, the thirty pages of 'Seasonal fertility rites and the death cult in Scandinavia and China' is quite short. Its composition is plain. After a short introduction, in which the aim of the work is presented, follow three sections on seasonal rites: 'Christmas as the festival of the living', 'Christmas as the festival of the dead' and 'Fertility rites, the cult of the dead and the life-promoting annual festivals in China'. The text ends with a section headed 'Christmas customs regarded as a means of protection'.

For an observer today, Andersson's complaints are rather difficult to understand. There are no objections regarding the structure of the text. The argumentation is logical and systematic. Hanna explains that the motive for her paper derives from the results of her previous article in the *Bulletin*. As summarised above, here her point in this was in line with Andersson's, that there was a difference between domestic and mortuary ceramics. Hanna extended this proposition not only to China but also to Scandinavian Neolithic and other areas, for example the Mediterranean regions and further east. She also reached the conclusion that the ornamentation of mortuary pottery had a dual symbolic meaning: life and death. These may seem incompatible, and to explore that alleged contradiction Hanna wrote this new article, turning to 'the ethnographical sphere' (Rydh, 1931: 69), focusing on folk memories of old customs as well as other ethnological, ethnographical and historical evidence. Within this bulk of material 'the two factors of life and death, the fertility rites and the rites for the dead have hitherto appeared as two irreconcilable contrarieties' (Rydh, 1931). From Hanna's perspective however, the ethnographic material contributed 'entirely new foundations' for the discussion, demonstrating 'the appropriation by the death cult of certain fertility rites' (Rydh, 1931).

The main point of the discussion was to explore the performative practices of various rites, and what meanings they conveyed regarding fertility and death. These meanings were based upon contemporary scholarship's notions of prehistoric, medieval and rural manners and the customs of 'the greatest festival of the North' (Rydh, 1931), the celebration of Christmas, or Yule – by using the latter term, the heathen origin of the festival was emphasised. This festival occurred on a seasonal basis every year, consisting of a series of events that followed a repeated, ritualistic pattern – reiterations in Hubert's sense – stretching over a specific period of time – Hubert's temporal milieu.

The pre-Christian Yule, according to Hanna's sources, was a festival connected to the return of the sun, the regeneration of fertility and the reassurance of prosperity. Ethnographic scholarship was presenting many proofs of such pre-Christian ideas and activities that, in a transformed or modified shape, had been integrated into the Christian celebration of the birth of Christ. Within the formalised sequence of the festivity weeks, Hanna highlighted many examples of performative events which she interpreted as belonging to the fertility cult. Examples were various practices that linked the harvest of the previous year to that of the coming year (Rydh, 1931: 73, 76–7), a preference for marriage within the period of the festival (Rydh, 1931: 75, 77, 81) and the pre-Christian toast to a prosperous and peaceful year (Rydh, 1931: 71) which was transformed into a toast to Christ and the Virgin Mary. The ethnographic evidence also indicated that there were close connections between some fertility metaphors of the Christmas festival and its solstice reverse, the annual festival at Midsummer, the festivity at which fecundity symbolism was supposed to be as most pronounced (Rydh, 1931: 72).

So far, the article provided arguments that the days of the Christmas festival were made up of a sequence of ritual events, aiming to promote fertility and life's regeneration. However, the rites and popular beliefs would also provide indications of the Christmas festival as a celebration of the dead.

In ancient rural society there were substantial beliefs that, during Christmas time, and in particular the night of Christmas Eve, ghosts and spirits were abroad, seeking contact with the world of the living. So, for example, there was a widespread belief that on the night of Christmas Eve the spirits of the dead would return to the church to celebrate mass before the early Christmas service (Rydh, 1931: 79). Some of these were understood as spirits of family ancestors and relatives who had recently passed away, eager to visit their home. Accordingly, a table was laid with food, surrounded by the festival's attributes associated with regeneration and growth. Even beds were prepared, waiting for the dead (Rydh, 1931: 78–9). Hanna's conclusion regarding these customs was that '[t]he dead returned in order to take part in the life-giving fertility rites that were to maintain life or regenerate them' (Rydh, 1931: 81).

In Hanna's opinion, a way to verify the hypothesis that 'Christmas which undoubtedly was a fertility feast at the same time was a feast for the dead' was to compare this dual notion and its attributes with another major festival to the dead: the feast of All Saints, immediately followed by the feast of All Souls. The latter was 'consecrated to the memory of the dead'. This festival, which was introduced by the church in the Middle Ages, occurs in the autumn, and according to Hanna would coincide with the existing, traditional harvest feasts of rural communities. She argued that there was a 'logical consistency in coupling the fertility feast with the feast of all souls' (Rydh, 1931: 84), thus finding evidence for the close and mutual connection between death rituals and fertility rites. It was in the interest of the Church to 'move' the souls of the ancestors and its pagan reminiscences (Rydh, 1931: 80) from the celebration of the Holy Birth, to the fertility feast in the autumn. By this, the connection between the souls of ancestors and harvest, fertility and regeneration was kept.

Bearing Hubert's theoretical perspective in mind, a close reading of 'Seasonal fertility' indicates that Hanna's approach to cult, myth, folk ethnography and folklore was of a similar kind as Hubert's, something of which the choice of seasonal festivities as the subject for analysis bears evidence. Within this theme, the recurring seasonal – and thereby temporal – performative practices, sequenced in ritual events, are in focus. The various examples of such events emphasise the temporal connection between rite and myth, linking to Hubert's analytical concept of temporal milieu. In line with Hubert's opinion, Hanna shows how a social constitutive principle – in her case the dual bond between death / ancestor cult and fertility cult – was permeating the period with its myths and rites in its entirety. As worldly time differs from mythical time, the temporal connection between myths and rites could also be spread out – this may be the reason for Hanna's proposition that, in spite of the temporal difference, there was a constituting unity between the death and fertility cult of the festivals of All Souls and Christmas.

However, as she makes no explicit references to Hubert, it cannot be proved that Hanna actually adopted his ideas from her meetings with him in person or from reading his work. But the assumption is highly probable, as Hubert´s main ideas were formulated prior to Hanna's stay at St Germaine, thus being accessible in the intellectual milieu. Many of the foundations of the Durkheimian school were explored as notes and reviews under various disciplinary sections or headings in the *Année Sociologique* and, as early as 1903 in volume 6, Hubert and Mauss jointly published an introduction and conclusion to the section of Myth. In this they outlined the main principles of the Durkheimian understanding of myths. Over the years Hubert wrote more texts about myth, and several important ones were published before Hanna's first visit (e.g. Hubert, 1905, 1919; Strenski, 1985).

As for many other European scholars, Hubert's academic work was interrupted by the war. This might have been one reason why, at his death, he 'left several unfinished works' (Isambert, 1983: 156), among them his life-long work on the prehistory and early history of the Celts and the Germans. As these studies were integrated into his lectures in archaeology and prehistoric religion at the École du Louvre and the École Pratique des Hautes Études, the result of his research was saved as drafts and notes. They were later compiled and edited by some of his colleagues, among them Mauss (Isambert, 1983: 156). The book about the Celts was published seven years after his death, with an English translation published the same year (Hubert, 1987 [1934]). The book about the Germans was delayed for another two decades (Isambert, 1983: 156). Thus Hanna could not have read them; however, at least some of the content had most likely existed as drafts when Hanna was in Paris – the book about the Germans is based on lectures given at the École du Louvre between 1924 and 1925, lectures that Hanna had planned to follow if her first visit had not been interrupted. Hanna most likely had the chance to discuss such matters with Hubert during her various stays.

The influence of Hubert's thinking is particularly visible in part three of *The Greatness and Decline of the Celts* (Hubert, 1987 [1934]). Here, the topic 'The civilization of the Celts' is introduced with the passage 'Objects and method of a sociological study of the Celts' (Hubert, 1987 [1934]: 185). This heading sets the agenda regarding Hubert's sociological perspective of prehistory and early history. It is conspicuous that this sociological perspective on the ancient Celts demonstrates many similarities with significant features of Hanna Rydh's study of Scandinavian seasonal festivities. One such similarity is that Celtic society, like the Scandinavian, demonstrates the basic relation of fertility rites and death cult. Hubert proves with many examples that Celtic religion and mythology were expressed through rites, and these myths and rituals were connected both to their beliefs in fruitfulness and life, and in soul, death and origin, i.e. ancestry (Hubert, 1987 [1934]: 235–6). Thus, one important feature was the death cult's connection to ancestor worship and ancestor regeneration. Just as in Hanna's text, the rituals were crucial for maintaining the religion and its constitutive myths. In particular the 'great seasonal feasts of agricultural life marked a momentary concentration' (Hubert, 1987 [1934]: 239) when 'the spirits got loose and wonders were expected and normally happened' (Hubert, 1987 [1934]: 240).

So even if we cannot prove a connection through explicit referencing in the text, we can see that some of the essential features in Hanna's discussion of religious folk life in the Scandinavia of earlier times coincide with fundamental elements in Hubert's understanding of Celtic religious life.

However, one very obvious connection to the Durkheimian historical ethnography as an explanatory device (Strenski, 1987: 360) is Hanna's use of the work of Marcel Granet as an authority. The scholarship of Marcel Granet plays an important role in the section 'Fertility rites, the cult of the dead and the life-promoting annual festivals in China' (Rydh, 1931: 86–96). Hanna explains this 'excursion to Chinese ground' with her view that it 'still further emphasises the important part played by the fertility rites in the cult of the dead' (Rydh, 1931: 96). This 'Chinese excursion' was based on J.J.M. de Groot's *The Religious Systems of China*, vol. I from 1892, but it relied even more on Marcel Granet's *La religion des Chinois* from 1922. From the turn of the nineteenth century and onwards de Groot (1854–1922), a Dutch scholar, Professor in Leiden and later Berlin had published major works on Chinese religion. As a young student Marcel Granet (1884–1940) came into contact with Mauss and the Durkheimian circle. While specialising in Chinese language, history and civilisation, he included the anthropological and sociological perspective, in which religion and myth were important parts. He was much appreciated by Mauss, and when Mauss in his biographical notes grieved over Hubert's untimely death, he wrote: '[b]ut we still have one mythologist, and that is Granet. There was one other, no less brilliant, and that was Hubert. I am trying to make up for the loss of Hubert and help Granet' (Isambert, 1983: 155, n. 11).

In the parts of the book Andersson refers to in his letter to Karlgren, Granet (1922) discusses the religious life of rural peasant communities during the feudal era in the first millennium BCE. Here are many examples of fundamental ideas that Hanna adopted while comparing Chinese conditions with the Scandinavian geographical and cultural area. Moreover, seasonal festivities in spring and autumn demonstrated the dual unity of the rites of worship linked to the cult of the ancestral spirits and the cult of fertility (Rydh, 1931: 64). Rituals connected to birth and burial linked the dead ancestor to the soil and the tilth in a 'manner that is of peculiar interest to us' (Rydh, 1931: 86), demonstrating – in Rydh's way of understanding – the constitutive unity between ancestral cult, regeneration and fertility.

Local critique

The critical letter from Andersson to Karlgren tells us that Hanna Rydh's first draft was met with a rebuff. As this version is not preserved, we cannot know what was changed. However, we can conclude that some alterations were made. For example, the title follows Andersson's suggestion. Hanna might also have shortened the text, as Andersson specifies her use of de Groot and Granet for seventeen pages, while in the finished article, the section on fertility rituals in China reached ten pages. Andersson further states that Hanna wrote a 'rather passable journalistic combination' of the two scholars. It is difficult to interpret the exact meaning of these words. Maybe the characterisation of the text as a journalistic piece of work indicates that Andersson felt it did not have a sufficiently scientific approach. As Granet's work does not provide a note apparatus – not needed in the series he wrote for – this lack may have affected Andersson's opinion. If the scientific quality of Hanna's first draft was questioned, we can see that measures must have been taken: in the printed version the structure is logical and well explained; she provides an account of the background to the work (Rydh, 1931: 69), a clarification of the research questions (Rydh, 1931: 69) and a formulation of hypothesis (Rydh, 1931: 70). Further the partial results (Rydh, 1931: 84) and main results are well concluded (Rydh, 1931: 86). This said, it does not mean that the text achieves all of today's *desiderata* for a scholarly product. For example, a more explicit awareness of the theoretical approach and a critical discussion of the explanatory value of the empirical evidence, which obviously derive from various cultural contexts, would have contributed to a more consistent product.

Another aspect of Andersson's slightly scornful journalistic attribution could have been the language. Maybe the first draft did not fit into the norms of a conventional scientific text, if it relied too heavily on the sources as regards language. Hanna Rydh herself characterises Marcel Granet's writing as a 'brief and highly interesting, popularly written presentation ... [in an] animated style' (Rydh, 1931: 86). Checking the text in Granet (1977), it seems that she uses many of his words (Rydh, 1931: 86–7) and two full pages are direct quotations from his book (Rydh, 1931: 87–9), obviously in her own translation from French to English. In this connection it might be interesting to note that regarding the English translation of Granet (1922) made by the esteemed anthropologist Marcel Freedman (Granet, 1975; here 1977), some reviewers have pointed out Granet's poetic sensibility (Wright, 1977: 696). In his translator's opinion, while using his prose he performed an act of almost 'scholarly prestidigitation' (Fried, 1977: 159), making this reviewer – Fried – regard Granet as a scholar, making 'airy leaps from poetic shard to greater than life sized reconstructions' thus being a 'weaver of gossamer webs' (Fried, 1977: 160). With this in mind, it is not difficult to see why Hanna was captured by such scholarship.

A bit peculiar is Andersson's disparaging description of Granet as 'little Granet'. Both Andersson, Karlgren and Granet were highly competent in Chinese language, culture, and history. The authority in Chinese studies at the time, Edouard Chavanne, was Granet's teacher and more or less his mentor. Karlgren, who studied in Paris for two years before the war, had also been a student of Chavanne. In the early 1920s Granet was one of the founders of the Institut des Hautes Études Chinoises in Paris, and when Andersson made his comment to Karlgren, Granet had for many years been professor and administrative leader of this research centre. So Granet was definitely Andersson's and Karlgren's peer regarding scholarship and social rank. It remains an open question why Andersson and Karlgren show this negative attitude towards him. Maybe the answer is as simple as that Karlgren and Granet sometimes had different views in scholarly matters. Some comments in Freedman's edited English translation of Granet (1922) hint at such circumstances (Granet, 1977 [1975]: 160, n. 11). Did Hanna involuntarily place herself at a point where two competing research networks intersected?

Whatever the reason for the letter, it criticises Hanna's article for scholarly quality, possibly for its textual character and probably for its theoretical, implicitly Durkheimian approach. As editor of the series, Andersson was within his rights to approve or disapprove of the contributions. Nevertheless, we can see that he uses tactics that we now recognise as techniques of dominance, an analytical tool that the Norwegian sociologist Berit Ås (1978) has developed to study processes of inequality. Within this male, homosocial arena, Hanna's professional competence was made invisible, ridiculed and belittled. However, with adjustments Hanna's paper could not be rejected – even if it was hard to 'swallow the pill'. To his colleague and fellow academic Andersson straightforwardly points at the decisive factor for accepting the paper: money.

We can see how Hanna accepted the compromises, but in her negotiating process she also applied a counterstrategy to stay true to her own scientific scholarship.[3] In her 1929 article for the *Bulletin*, she had stated that the protective significance of some rites was of secondary importance to the fertility notions. Andersson raised the objection that she had not acknowledged his view of the prophylactic importance of certain rituals during the seasonal festivals. In the article of 1931, Hanna came back to this issue (Rydh, 1931: 96–8), once again questioning Andersson's notion of the prophylactic primacy. She saw it as 'a problem subsidiary to the Christmas customs' even if by others it were 'considered as a problem of the first importance' (Rydh, 1931: 96). Presenting a number of pieces of empirical evidence that in her opinion were persuasive, she acknowledged the relation between fertility and protective significance; however, 'I will only repeat that I am convinced that the protective significance is secondary' (Rydh, 1931: 96). By developing a different opinion, Hanna thereby placed herself in the position of a professional equal.

Discussion and conclusion

By following the genesis and content of Hanna Rydh's articles we can see how specific formal and informal networks were crucial in the process of producing the knowledge in them.

The surrounding Swedish professional community supported Hanna in completing her doctorate. An advantage was that her own husband was part of this encouraging group, being a partner in discussions and bringing access to specialist knowledge. Having passed the threshold of graduation, parts of her research followed the general line of research, something of which her male colleagues generally approved. The professional archaeological network also accepted her popular texts with female focus which were mainly addressed to women. However, when Hanna diverged from the conventional scholarly track by introducing new scientific research questions, as she did with her second article in the *Bulletin*, the professional community was less supportive, even resorting to what we now understand as techniques of dominance. Obviously, Hanna's attempts to get access to the circle around the East Asian Museum were conditional. The circle around the East Asian Museum seems to have been a very masculine, hierarchical and homosocial milieu; when the fourth volume of the *Bulletin* was dedicated to the Crown Prince as a token of gratitude for his 'careful studies' and 'dedicated interest... [in] the old art of the Far East' sponsors could sign a tribute. Of the ninety-one persons who signed this list, only five were women (*Bulletin* no. 4, 1932: V–VIII).[4]

The academic female network played a most significant role. Female academics informed Hanna about the international scholarship, encouraged her to apply and sponsored her international studies financially. This financial support made it possible for Hanna to get access to a scholarly arena which, in the early 1920s, she considered most likely to bear fruit for her professional development.

We can see that Hanna embraced the feminist movement's international network with a dual approach: both as a receiver of support for knowledge-producing purposes and as a supporter of the expanding women's educational 'room'. This underlines that within the feminist circles of the 1920s education was seen to be of vital importance for female emancipation. It is also significant that the means for the fund-raising campaign demonstrated in this chapter was domestic handicraft, i.e. primarily textile objects that were made – and controlled – by women.

Returning to Berit Ås' observations on techniques of dominance, these analyses have resulted in the articulation of several counterstrategies, such as the spreading of information, supportive encouragement and providing venues in which the underprivileged subject can be visible and get a place in the social 'room'. All these traits can be observed in the strategies used by the IFUW network.

It is also possible to regard the particular objects that constituted the gift of domestic handicraft as secondary agents or actants. In accordance with ANT they can even be understood as constituting a particular space-related network in their own right. They were produced in various homes in Swedish rural areas, gathered at local museums and forwarded to a collecting place in Stockholm. After being valued and given a price they were shipped to London, ending up at a bazaar and eventually transformed to economic means for creating a room for the production of knowledge. Infrastructural means for these movements were private bicycles and local bus lines, national railways and international, North Sea, shipping lines, thereby giving a sample card of the geographic technology of the time.

The archaeological research milieu at St-Germaine-en-Laye where Hubert was the central figure was welcoming and encouraging to Hanna. Her gender does not seem to have posed any problem. In spite of several interruptions, she was given both physical and affectional room to finish her research and develop her studies. It is also evident that she met a research milieu, the Durkheimian social historical approach, which had not yet arrived in Sweden, in which the cultural history approach was dominant (Baudou, 2004). Even though the Durkheimian school constituted an established intellectual domain (Besnard, 1983), for Hanna it was a loose network, related to personal communication, possibilities to attend lectures and get access to libraries and archaeological collections. The archival documents do not reveal the nature and extent of the French connections; however, a close reading of Hanna's work in which she explains her reasoning and scientific inspiration, indicates that she was influenced by the Durkheimian ethno-historical sociological approach. It is also interesting to see that in spite of being inspired by Granet, she chose to write her text in English, even though the *Bulletin* also welcomed articles in French and German, both languages that in the Swedish academy were more conventional for scholarship.

Hanna Rydh's family constituted a fundamental supportive network. Bror Schnittger shared his knowledge of the scholarly community so Hanna could get access to a type of knowledge which was an asset while planning her scientific development. Schnittger also supported Hanna's wish to combine a professional career with raising a family. Regarding Hanna's father's support to the Museum of Far Eastern Studies on 'condition' that Hanna could participate in the Museum's research, this was a far from unique model. The historiography of archaeology shows several examples of how families of fortune supported cultural or scientific projects, where their daughters were appointed to perform a specific task, giving them access to a profession (Arwill-Nordbladh, 2008: 160). As a conclusion concerning the importance of the familial network in these instances, the conventional divide between private and public seems to be more complex than generally assumed.

The geography of knowledge production and the knowledge-producing 'rooms' in Hanna Rydh's networks varied. The desk in an office at Saint-Germaine-en-Laye and the sorting table for pottery at the Museum of Far Eastern Antiquities designate the museum as an important site for knowledge production. Auxiliary institutions like archives, libraries and lecture rooms were also such arenas. Here Hanna, being a female archaeologist in an all-male context, could relate to and navigate within the conventional social structure, occasionally aided by members of her family circle.

The newly established study house for female university students is an example of how a determined group of feminists were stretching the normative gender structures, promoting women's access to scientific education and thus creating new scholarly 'rooms'. In the long term, this affected the conditions for producing knowledge, challenging the conventional masculine approach of academia.

It is noteworthy that Hanna Rydh's knowledge-producing clusters – including their material and literary character – were geographically and temporally widespread. London, Paris, Stockholm, rural China,

Celtic Gaul and the network of Swedish countryside homes in which domestic handicraft was manufactured, were some of the nodes that converged in Hanna Rydh's practices. Creative milieus for international feminist emancipation, a Durkheimian theoretical hub and a museum with a touch of male supremacy based on collections of exquisite ancient art, constituted intellectual clusters and structured sites. While moving between them as an agential subject with a vision and a goal, Hanna Rydh shaped her scholarly products.

As to whether structure or individual agency has priority in the shaping of life, Hanna Rydh's achievements show that her integrity, determination and visions for the future guided her lines of conduct and helped her to seek and choose while manœuvring within and between various structured networks.

Epilogue

As mentioned earlier, Hanna's two articles in the *Bulletin* do not seem to have had much influence upon the contemporary Swedish archaeological research community. Few, if any, references can be noticed in other scholarly works. It is possible that Hanna's approach was not in line with the conventional strand. Nor did Hanna herself continue down this newly trodden path. In August 1929 she wrote to her friend the Danish runologist Lis Jacobsen that her life was going to take 'an unexpected turn' (letter from Hanna Rydh to Lis Jacobsen, 6 August 1929, The Royal Library, Copenhagen). While the first article was going to print, she married her second husband, Morimer Munch af Rosenschöld (1873–1942), undersecretary of state in the Ministry of Education and Ecclesiastical Affairs, who in 1931 was appointed county governor of Jämtland and Härjedalen. Her professional task had to adjust to her obligations as the wife of a county governor (Lundström, 2005; Arwill-Nordbladh, 2013). The articles about death cults and fertility and the Durkheimian inspiration to a historical-social approach to prehistory can be seen as solitaires with little impact on Swedish archaeology. However, they can also be seen as pieces of scholarly work made by an intellectual and in many ways original mind which was open for negotiating within and between different networks. These processes illuminate the contested nature of the production of scientific knowledge.

Notes

1 Sondheimer, J.H. 1958. *History of the British Federation of University Women, 1907–1957*. London: BFUW: 39, in Arwill-Nordbladh, 2005a: 133.

2 Gothenburg University Library, Kvinnohistoriska samlingarna A 12, Hanna Rydhs samling I: 15.
3 See for example, *Bekräftartekniker och motstrategier* (affirmative techniques and counterstrategies): www.jamstallt.se/docs/ENSU%20bekraftartekniker.pdf.
4 When a signature only includes initials and surname, I have placed it in the male group. It may be interesting to note that both Hanna Rydh and her brother C.L. Rydh were among the signers. Hanna's father had by this time passed away.

9

'Trying desperately to make myself an Egyptologist': James Breasted's early scientific network

Kathleen Sheppard

Introduction

On Tuesday 30 October 1894, James Henry Breasted (JHB) wrote to his parents back in Rockford, Illinois from his steamship in the Mediterranean: 'Just think of it! I am within a few hours of the shores of Egypt and will soon be among the scenes I have studied so long. It seems hardly credible. Now I hope to use every moment and hasten back to my homeland and all I love as soon as ever I can' (JHB Papers, Box 4).[1] Breasted's first expedition to Egypt in 1894 as a newly minted Egyptology PhD would be crucial to his career – and he knew it. Not only would the journey provide him with experience in the field, which he needed in order to be considered a true professional Egyptologist, but it would also allow him to build the dynamic scientific network that would aid and sustain his work within Egypt for the next forty years. Scientific networks are essential to the practice of science, both in the field and within institutions. Furthermore, investigating the locations in which scientific networks are formed is pivotal to how these groups interacted within themselves and among other networks, as well as affecting what kind of knowledge they produce, if any. Place is crucial in the study of the development of scientific networks and the manners in which scientists communicate.

Studying the groups of colleagues, assistants, students, and staff is not new to the history of science, but in the case of archaeology and Breasted's career it bears some explicit discussion here. Scholars who study present-day scientific networks argue that the best way to visualise their connections is through tracing joint publication and reviews of those publications (e.g. Newman, 2001; Glänzel and Schubert, 2005).

Historians agree with this assessment but recognise the limitations of this medium, arguing that correspondence is a key piece of evidence in understanding how networks interacted with each other outside of publications, that is, out of the public eye (e.g. Secord, 2000; Finnegan, 2005; Fyfe and Lightman, 2007; Browne, 2014; Sheppard, 2018). In archaeology especially, the main groups of scholars who influence each other tend to gather in the field, in ephemeral groups among which some members are permanent fixtures every dig season, others come and go, and still others only appear once, briefly, and then vanish into the dust of the site and archive. Their connections do not necessarily appear in joint publications and are, therefore, hard to trace. As Janet Browne has argued, studying correspondence among scientific networks allows 'the prospect of reconstructing patterns of sociability with due appreciation to the structure of the society in which they emerged' (2014: 169). We then gain insight into schools of thought, in order to better understand who is participating, who is allowed to share ideas, and how those ideas are shared.

Further, we must understand the places in which knowledge is being created. Just as David Livingstone (2003: 13) has argued that 'scientific knowledge bears the imprint of its location', I argue so too do collegial relationships. Where science is done depends on who is able to, or allowed to, participate in the creation of knowledge; the reverse is also true, that is, who is allowed to create knowledge depends on where science is done (e.g. Naylor, 2002, 2005; Livingstone, 2007; Livingstone and Withers, 2011; Terrall, 2014). Geography of knowledge determines how relationships within scientific networks operate depending on where they were built, where they operate, and where and how their knowledge is spread. To comprehend this, it is crucial to understand who is interacting at different types of site such as universities, excavation sites, museum offices, private homes, hotel dining rooms, and formal scholarly meetings.

In Egyptology, knowledge is created, discussed, and refined in every space from the university or museum office in a disciplinary center out to the unceremonious field site and back. In early Egyptology, network hubs tended to be in metropolitan cities all over the world. Societies like the Egypt Exploration Fund (now Society, in London), institutions such as the Cairo Museum, British Museum (London), the Oriental Institute (Chicago), Metropolitan Museum of Art (New York), the Louvre (Paris), the Museo Egizio (Turin), and the Berlin Museum were all hubs from which and to which scholars and their ideas traveled. As the formal institutions in Egyptology, these centers retain records of membership, meeting minutes, details of official activities and decisions, and formal collections of scholarship. The records held by these institutions help to tell a rich story of discipline formation from an official point of view. Their published collections are crucial to historians and other scholars who trace changing ideas over time. Through some collected correspondence, historians are also able to view the overlapping members at each of these metropolitan hubs. Senders and receivers of letters, as well as the subjects of those letters, are central to understanding the story of who was present and active in multiple networks, what they said to each other, and how they interacted. In the field, that is, outside of metropolitan institutions, these developments and movements are harder to trace. But they are the mundane everyday activities that ended up being even more central to the formation of the professional discipline than published scholarship and institutional organization.

This chapter focuses on James Breasted's early professional network, specifically the two nodes that he cultivated on his first trip to Egypt: the British field archaeologist Flinders Petrie and the French Director of the Department of Antiquities in Egypt, Gaston Maspero. These personal and professional networks then expanded from the institutional hubs into the broader scientific discipline of Egyptology. In scientific networks, nodes are the people around whom subnetworks can and do form. The people in those subnetworks and the relationships among them dictate what kind of professional or personal activities happen. Using Breasted and his relationships with Petrie and Maspero as brief case studies, I will examine the importance of place in building and maintaining scientific networks for the field scientist. My contention is that scientific relationships built primarily at an isolated excavation site – a space far removed from conventional institutional settings – made relationships informal and familiar, much like the field itself. On the other hand, connections developed in urban areas or within formal and established scientific institutions, such as in universities or museums, tended to maintain that decorum, as well as be reflected in the types of work the scientists do together. I reveal the nuances behind these varying sites of knowledge creation and the effect that the rural field site Petrie occupied or the urban institution Maspero led can have on the development of scientific networks. While detailing each of these instances would be a book-length study, Breasted's example will illustrate the point and hopefully lead to further discussion by other scholars. For Breasted's early scientific network, the informal excavation site at Naqada as well as the metropolitan centers at the University of Paris and the Cairo Museum were hubs of knowledge creation and forging the bonds of scientific relationships. Each location produced different evidence, different relationships, and different scholarly outcomes.

Correspondence and field diaries, usually found in archives at the metropolitan centers, are essential to tracing network participation in different areas of the scientific practice. Scholars are able to witness activity described within the correspondence at these institutions, and it is fascinating to see how, when, and where people moved throughout the excavation seasons and their off-seasons. Using this evidence, we are able to trace the creation of scientific knowledge within, outside, and throughout the hubs and the field. The evidence I use here is taken mainly from archival correspondence and published biographical accounts of the characters. Even with giants of the field like Maspero, Petrie, and Breasted, it is difficult to trace personal and professional relationships through the literature. Therefore, my conclusions are based on anecdotes from their published life stories as well as a theoretical appraisal of correspondence as biography (C. Breasted, 1943; Drower, 1985; Abt, 2011). James Breasted is the central character of this particular investigation for two reasons. First, he is largely recognised as the earliest university-trained American Egyptologist, which meant that his career trajectory would be vastly different from European Egyptologists at the time. Second, because Breasted's career would set the foundations of academic Egyptology within the United States, he knew from the start that it would be necessary to form his scientific network carefully and deliberately. His case therefore allows for the explicit examination of strategic network building.

James Breasted, Egyptologist

James Henry Breasted was born on the prairies of Illinois in 1865. By the time he was 22, he had shifted careers from pharmacy to the ministry, which he pursued at the Chicago Theological Seminary. While there, his Hebrew professor Samuel Curtiss found Breasted's linguistic ability to be a useful asset. He convinced Breasted to continue studying Hebrew and to pursue a career in what was then a 'vacant field' in America: Egyptology (Abt, 2011: 6–19). Breasted went first to Yale to study Semitic languages under William Rainey Harper, receiving his master's degree in 1892 (Abt, 2011: 23). Following Harper's advice, he began studies at the University of Berlin in 1891 in Egyptology with one of the foremost scholars of the day, Adolf Erman (Abt, 2011: 19–26). After three years of intense study with Erman, a brutal exam process, and a dissertation written in German, then translated and hand-written in Latin for publication, Breasted earned his doctoral degree. Although he had completed all the university requirements to earn the title *Herr Doktor der Philosophie* Breasted, *Hochwohlgeborner*, and had a job waiting for him at the new University of Chicago, he still had to make the journey to and through Egypt to establish himself as a true professional.

As with many field sciences, a degree in Egyptology alone did not give Breasted professional standing. Erman thus urged Breasted to go to Egypt 'for the sake of his health and scientific future,' and gave him an important task: collating inscriptions in the Egyptian Museum in Cairo for a massive dictionary Erman was writing (C. Breasted, 1943: 51). Understanding the importance of this fieldwork, Breasted scraped together money from a variety of sources. Writing to Harper, his old professor at Yale who was to be the new President of the University of Chicago, he argued that he would not only be able to get objects for the new Haskell Oriental Museum, but he would also gain essential practical knowledge on such a trip. Eventually, Harper allowed Breasted a fully paid leave from his position at Chicago before his duties even began, and his parents were able to give him money for the journey, amid their own hardship. He reassured them: 'Apart from its usefulness for my studies, the Egyptian trip would be a replenishing of the man, and a lifelong inspiration. It would be a godsend before settling down to the grind at the University of Chicago' (quoted in C. Breasted, 1943: 52). Breasted was soon on his way to Egypt, with his new wife Frances. As he understood and had explained to his family and Harper in the United States, this first journey to Egypt was crucial to his professional status and the development of his network of associates in Egyptology.

Breasted: Petrie's Pup

The Breasteds first arrived in Cairo in early November 1894, and took inexpensive lodgings as they prepared both to work and to celebrate their honeymoon. While getting their supplies organised for a two-month trip down the Nile, they met and forged friendships with already-famous archaeologists, such as Archibald Sayce and Jacques de Morgan, and a number of British and American tourists. Breasted spent every spare minute working at the Egyptian Museum on Erman's dictionary project and completing some tasks for Harper at Chicago. Harper had written to inform Breasted that he had been voted by the Board of Trustees to be a 'representative of the University to receive gifts.' On top of being paid his full salary throughout the trip, Breasted was given $500 for the purchase of Egyptological photographs, casts and artifacts for the building of the new university's Haskell Oriental Museum collections. Harper encouraged him in all of this, writing: 'This I think will show you our appreciation of the situation and will redeem in part at least the pledges given you for assistance this year… I hope you will do your best to secure material, having in mind especially the practicability of the material for us, and not mere curiosities.' He then wished Breasted a 'satisfactory... and most successful trip' (September 26, 1894: JHB Office Files, 1894–6).

These activities were no doubt the start to making Breasted a 'real' Egyptologist. By meeting with him and encouraging his work, Erman, Sayce, and DeMorgan supported these activities as legitimizing Breasted as a true field scientist. Contributing to Erman's project as well as acquiring objects for the Museum, Breasted had his work cut out for him. He was employed by the top linguist in his field and his new university was entrusting him to build their Museum collection from the ground up. Further, that he was doing these activities in Cairo made his case as a professional Egyptologist even stronger. But he was still missing a key component.

Knowing full well he would be busy in Egypt trying to become an Egyptologist, before he left Berlin Breasted had taken the first step in building his network so that he would indeed have a successful trip. He sent copies of his thesis, which analyzed ancient Egyptian hymns to the Sun under Amenhotep IV (Akhenaten), to a number of archaeologists and Egyptologists in Europe. He received favorable responses from many of them. The most exciting response Breasted received was from University College London's Flinders Petrie. Petrie was a self-taught British archaeologist; although he lacked formal education, steady funding, and experience, he began his work in Egypt in the early 1880s on the Giza plateau, surveying the pyramids (Drower, 1985: 34–64). Petrie quickly established himself as an authority in the field and by 1892 held the first university chair of Egyptology in Britain. Even more quickly, he established himself as an infamously frugal excavator. By the time Breasted met him in 1894, Petrie was undoubtedly the leader in excavation practices in Egypt. He had a ready group of students who he, Walter Crum, and, later, Margaret Murray, trained in the classroom at UCL and from whom he had his choice of field assistants, known as Petrie's Pups (Janssen, 1992: 12–13). Aside from acknowledging Breasted's thesis in his response, the Professor also offered 'some kindly advice, the promise of some things for our Museum and above all an invitation to come & spend some time with him at his excavations of Coptos!!!!' (November 1, 1894, JHB Papers, Box 4). Upon his arrival, Breasted found a ready mentor and Petrie had a willing new Pup.

In his letter, not surprisingly, Petrie also advised the newlyweds on the cheapest way to travel in Egypt: by boat on the Nile (C. Breasted, 1943: 64). In accordance with Petrie's advice, and through the help of an Egyptian acquaintance, the Breasteds ordered a *dahabeya*, or houseboat, on which they would sail down the river from Aswan to Cairo. The Breasteds reached Asyut by train from Cairo, sailed up the Nile, South, to Aswan first in order to then sail down the river, North, with the current. They visited a number of sites along the way, including Luxor twice and Elephantine Island. The highlight of the trip for Breasted was getting to the Petrie camp in late December. It was such an important event for him, he later recalled, that upon reaching Naqada, Breasted 'jumped ashore and without waiting for a donkey to ride, hurried off on foot to find Petrie. His eagerness and the warm welcome he received made him oblivious to the long, tiring walk.' Breasted found Petrie on site, dressed, 'not merely careless but deliberately slovenly and dirty. He was thoroughly unkempt, clad in ragged, dirty shirt and trousers, worn-out sandals and no socks' (C. Breasted, 1943: 75). Despite his appearance, Petrie was ever the professional excavator, and trained Breasted in his meticulous methods. Petrie and his assistant James Quibell (one of the earliest Petrie Pups) were already ensconced in the site, excavating a large pre-dynastic cemetery. Because of the massive number of burials – they excavated over 2,200 graves in one season – Petrie had to quickly develop an organised system of uncovering and safely cleaning the grave, recording the finds, and collecting the objects.

He recorded this system in the field report for the season in order to 'give sufficient confidence in the general accuracy of the results noted' (Petrie and Quibell, 1896: ix). He explained that he used a 'compound gang' of pairs of Egyptian men and boys, led by his *reis* Ali Suefi (Quirke, 2010). He described the process:

> First a pair of boys were set to try for a grave, and if the ground was soft they were to clear around up to the edges of the filling, but not to go more than a couple of feet down. At that point they were turned out to try for another, and an inferior man and boy came in to clear the earth until they touched pottery or bones in more than one place. They then turned out to follow where the boys were working, and the pair of superior men came in to dig, or to scrape out with potsherds, the earth between the jars. While they were at work Ali was in the hole with them, finishing the scraping out with a potsherd or with his hands, his orders being to remove every scrap of loose earth that he could without shifting or disturbing any objects. When he had a favourable place his clearing was a triumph; every jar would be left standing, still bedded to the side of the grave, while all the earth was raked out between one jar and another; the skeleton would be left with every bone in its articulations, lying as if just placed on the ground, the cage of ribs emptied, and the only supports being little lumps of earth left at the joints. The flint knives or other valuables would be each covered with a potsherd, to keep it from being shifted and a pebble laid on that, to denote that it marked an object (Petrie and Quibell, 1896: viii–ix).

Petrie would then come to the grave to record the locations of the objects as well as finish the removal of the grave goods and skeleton.

His excavation and recording techniques were routinely detailed and focused on small objects such as potsherds and necklace beads, giving him the nickname 'Abu Bagousheh – Father of Pots' (see Stevenson, 2015).

Breasted spent almost two weeks with Petrie on the site and 'absorbed every detail of the technique of excavation, its supervision and cost,' as a good Pup was expected to do. Not only did he witness Petrie's new cemetery excavation technique in detail, but he also learned that, the previous year, 'Petrie had paid "just five shillings a week for provisions for himself and his assistant"' (C. Breasted, 1943: 76). Before leaving Naqada, Petrie had suggested to Breasted that they should collaborate on an excavation site for a season, and Breasted was happy to consider it. As an American, Breasted had the potential to bring a lot of private money with him, which the British were not able to secure (see e.g. Reid, 2015: 19–29). However, while Breasted believed that excavation was 'eminently worth-while,' it was only of 'secondary importance' to him. Instead, he 'foresaw that his own most important work in Egypt would be the reconstruction of her ancient past rather than the recovery of the material remains of her civilization' (C. Breasted, 1943: 77). This reconstruction project soon became his life-long goal, institutionalised in The Epigraphic Survey at the University of Chicago, whose continuing mission is to record all surviving inscriptions on temple and tomb walls in Egypt and publish them before they perish with time (Abt, 2011: 46–7, 281–301; Epigraphic Survey, 2014). Petrie not only taught Breasted how to excavate, but also allowed Breasted to realise that excavation was not his passion or purpose.

Throughout his career, Breasted continued to meet Petrie in the field and, later, in London. They never published any joint scholarship, so tracing their relationship through that medium would not reveal what the archival evidence does. For the rest of their long, collegial relationship, Petrie and Breasted exchanged letters about excavations, the Egyptian Research Account (ERA), and the British School of Archaeology in Egypt (BSAE) – both institutions run by Petrie at their early stages, and both institutions that gave objects to Breasted's museum. The two men would meet in London when Breasted was there, usually at the College (UCL), which was the headquarters for the ERA and BSAE, at the Petries' house, or out for a meal. Petrie did not keep a journal and Breasted did not write about the details of their meetings, but it can be assumed that in their 'talking shop' they spoke in person as they did in their letters – about excavations, objects, money, and sometimes family. They had much to discuss, in terms of what should be excavated and/or recorded as both Petrie and Breasted continued in their chosen lines of work.

They corresponded about their scholarship, each helping the other with translations, transliterations, and general editing. By late 1896, Petrie received corrections and editing assistance from Breasted for the first volume of his *History of Egypt*, which came out in six volumes over eleven years (Petrie, 1896; Petrie to JHB, December 29, 1896: JHB Office Files, 1894–6). Petrie also continued to depend upon Breasted and his connections for financial assistance. As early as February 1896, Breasted had committed to sending money to Petrie for the ERA's work in the field. By October of that year, Breasted had sent Petrie $155 for excavations, thus guaranteeing the Haskell Museum a number of objects from that season's work (Petrie to JHB, October 31, 1896: JHB Office Files, 1894–6). It is clear from private correspondence that they respected and admired one another and that they both highly prized fieldwork. Their collegial relationship would continue for the rest of their careers, and they remained warm friends as well.

Much as it had done, and would do, for generations of diggers trained by Petrie, Breasted's time on site made him a professional Egyptologist possibly more than his doctorate did. Petrie's goal in training excavators was to instill in them his methods of scientific archaeology. He was wedded to measurement, quantification, and careful extraction from the ground of as many artifacts as possible. While a PhD was important for the language study that Breasted wished to do, in order to be a real Egyptologist at the turn of the twentieth century he needed field experience and Petrie was the pre-eminent field Egyptologist of his generation. In order for people like Breasted to get the right field training, support from established professionals such as Petrie was crucial. By becoming one of Petrie's Pups, Breasted achieved the necessary training, created this particular node with Petrie at the center, and therefore gained acceptance into the well-established network of field archaeologists who had been trained by Petrie. Breasted's deliberate construction of this part of his network would support him for the rest of his field career.

Breasted and Maspero

With Gaston Maspero, however, the circumstances in creating this node, and therefore the activities that took place within it, were substantially different than with Petrie. As an established scholar and former director of the Department of Antiquities, Maspero outranked Breasted in the profession, so their relationship was undoubtedly marked by that disparity, at least in the beginning. Further, where Breasted and Petrie were field scientists, Maspero was a scholar and administrator, concerned with publications, schedules, permit approvals, and budgets. Their relationship reflected this in their correspondence, as well as their friendship. Breasted was aware of the dynamic when he met Maspero, which meant that this node and the network associated with it took on a markedly different tone than the node that Petrie occupied.

After a two-month Nile journey, complete with the two-week stop to see Petrie, the Breasteds' honeymoon came to an end. In early 1895, Frances and James made their way back to Chicago by passing through Paris and London to finish a few tasks. In Paris, Breasted visited the Louvre for the first time and worked in the Egyptian collections, copying, reading, and getting more experience in the discipline, while trying to professionalise himself. He had established himself in Petrie's network as an official Pup, but he needed training in a new space. He wrote that, during that busy week,

> I did spend an hour with Frances among the Asiatic collections, and ten minutes among the Greek marbles, to see the Venus de Milo. But I saw nothing of Paris and its environs, I learned almost nothing of the French, and moved like a mole through the wintry streets between a shabby little hotel and the Louvre. This was obviously not the way to broaden one's horizon or enrich one's cultural experience. It was, in fact, reprehensible and stupid. But I was trying desperately to make myself an Egyptologist according to a concept I had evolved alone and could not find words to impart to those around me (quoted in C. Breasted, 1943: 82).

Despite his time in Egypt, Breasted knew he still had work to do to 'make' himself an Egyptologist. This included study time in the controlled museum space, as well as continuing to build his network of support from other Egyptologists. Specifically, Breasted went to meet with 'the great Gaston Maspero,' who, he told his father, 'scientifically stands in France where Erman does in Germany'; high praise indeed from a German-trained Egyptologist (C. Breasted, 1943: 82).

Like Breasted, Maspero was as a linguist; he had worked with early experts Auguste Mariette and Heinrich Brugsch. He also performed numerous field duties as the Director of the Bulaq Museum and Antiquities Service from 1881–86 and again from 1899–1914. He was responsible not only for translating a number of now-famous Egyptian texts, but also for opening a number of small pyramids and other tombs, removing the Deir el-Bahari cache of royal mummies, and unwrapping some of those mummies for study (Bierbrier, 2012: 359–61). In order to prepare for the excavation seasons, archaeologists had to visit the Director in order to obtain permits to work in particular areas. Maspero was known as an agreeable Director, so in 1886 when he resigned his post some European archaeologists were distressed. In fact, many archaeologists found the French administration of the Department of Antiquities to be troublesome most of the time. Breasted had had trouble at the start of his trip in 1894, dealing with then-Director Jacques de Morgan. Breasted wrote: 'I find the administration of the antiquities… by the French, corrupt to the core. Nothing is done in the name of truth or science, but all is a mere scramble for good things to sell & the money goes into private pockets' (quoted in Abt, 2011: 44). The only Frenchman who seemed palatable to British, French, Germans, and Americans alike was Gaston Maspero. He wielded a great amount of power in the discipline, but tended to be friendly and fair, usually granting permissions for digging as requested. Most people found it difficult to dislike him (Hankey, 2001: 131).

Maspero had left his post in Cairo in 1886 in order to go back to Paris and take up new duties as Egyptology professor at the Collège de France; Breasted met him in his office there in 1895. Breasted wrote to his father about the meeting, telling him '[Maspero] received me cordially, talked delightfully for more than an hour about his books, his purposes, his youth and his present researches… He was kind enough to ask about my own work, graciously giving me the opportunity for presenting him with a copy of my Berlin dissertation' (quoted in C. Breasted, 1943: 82). This first meeting, although short, was the beginning of a scientific friendship between the newly branded PhD and the older, seasoned scholar and professor. Interestingly, soon after meeting Maspero in person, Breasted published a critical review of his *The Struggle of the Nations*, which did Breasted no favors with other French archaeologists but seemed not to have troubled Maspero (Abt, 2011: 51; see Maspero, 1896; Breasted, 1897). However, much like his opinion of de Morgan, his review of Maspero's work was doubtless impacted by his German opinions of French Egyptology (see Abt, 2011: 51).

Although both men were scholars – and not primarily excavators – coupled with the differences in their ages and the divide in their professional statuses, the friendship had been established in a formal place – the Great Man's professional, institutional academic office – and the relationship continued this formal pattern. Probably because of this formality, there were few letters between the two scholars when Breasted was not preparing to go to Egypt. However, being allowed into Maspero's network by establishing him as a node in his own network would open up a whole new scientific world for Breasted in terms of work possibilities and travel throughout Egypt. Breasted only needed to contact Maspero about official work, so in 1905, in preparation for his first trip to Egypt in almost a decade, he did.

For this trip, Breasted brought his family with him, consisting of his son Charles, then eight years old, and wife Frances. Before the expedition left Chicago in October, Breasted wrote to Maspero to ask permission to 'photograph or copy all the inscribed ancient monuments of Upper Egypt' (C. Breasted, 1943: 146). The Herculean request betrayed Breasted's lack of field experience and the process necessary for completing the job. Maspero, having taken up his position in Cairo again, responded to Breasted's request that the area he wished to cover was too vast and the monuments too numerous for Maspero to be able to grant the requested permission. Instead, he advised Breasted to 'select a special district in Nubia and to work it out before asking for a second one' (Maspero to JHB, October 16, 1905: JHB Office Files, 1905). He also told Breasted not to clean or excavate any of the sites. This was good advice from a seasoned field mentor. Upon their arrival in Cairo, like everyone else, the Breasteds needed to go in person to secure their permits. James took his young son Charles with him to meet Maspero in his museum office, which was distinctly different from the Paris University office where they had first met. Charles later described the setting as 'crowded with open boxes of recently excavated antiquities... Here, like treasure in a cave, was everything imaginable, all the things my father had told me I myself could perhaps dig up with a small shovel!' He remembered little of the meeting except that his father and Maspero 'talked their shop,' and that the permissions were granted (C. Breasted, 1943: 147; Abt, 2011: 126–53). Meeting in the museum office was not only necessary for fieldwork, but it was also a symptom of the formal relationship the two scholars had maintained. On both occasions they had met in Maspero's institutional offices; there is no known record of them sharing meals, teas, or meeting in other, informal places. But by 1905, Breasted had risen through the academic ranks in the United States and had earned his professional bona fides. He was the Director of the Haskell Oriental Museum, held the chair of Egyptology and Oriental History as full professor at the University of Chicago, and had just been awarded a generous grant for his work by the John D. Rockefeller Foundation. At this point, it would have been more a meeting of equals.

Outside of these two meetings, ten years apart, not much is known of Breasted's relationship with Maspero and the few letters between them mostly deal with releasing artifacts, getting permits, and a few other professional concerns. We do know that, unlike his relationship with Petrie, Breasted and Maspero only corresponded about work (although, as with many of his correspondents, Maspero often included some 'personal, even fatherly touch' in his letters to Breasted) (Hankey, 2001: 132). Further, their correspondence differed greatly from those letters Breasted exchanged with Petrie. We know that they met at the start of each season that Breasted went out to Egypt, until Maspero again left Egypt in 1914 to return to France at the start of the First World War. However, when Maspero's son died in the War in 1915, Breasted wrote to Maspero expressing sympathy for his loss (Abt, 2011: 220).

They reviewed each other's scholarship, as they had similar goals for academic Egyptology, but their respective concerns within Egypt were slightly different: they both wished to preserve what was there, but Maspero did so through promoting excavation where Breasted did so by recording and publishing.

Conclusion: Dr Breasted, professional Egyptologist

James and Frances returned to the United States in 1895, after their crucial first journey which took them through Egypt, Paris, and London, building Breasted's experience base and his scientific network, and, in turn, truly making him a professional Egyptologist. These colleagues formed the two foundational nodes in his early scientific network on which Breasted built the rest of his career as a professional Egyptologist. The Petrie node began as an informal mentorship and soon morphed into a partnership of equals. Others in this node were Quibell, Sayce, Francis Griffith, Walter Crum, and George Reisner. The node that Maspero occupied was a formal, foundational connection for Breasted's access to the field in Egypt, which included other museum personnel such as Jacques de Morgan and, later, Reginald Engelbach. Their letters revealed their in-person relationship dynamics: Petrie was a colleague and friend; Maspero was a formal mentor and official contact. Petrie had trained him to run a site and continued to be a friend, colleague, and practical mentor; he also provided objects for the Haskell Museum. Maspero had given him one of his first formal connections, permissions, and the confidence and advice he needed to pursue his goals in Egypt for the Epigraphic Survey. In each of their publications, these three men cite each other as scholars, but there are no joint publications among any of them. Therefore, tracing their relationships through correspondence is crucial to seeing the scientific network Breasted was building and how he was building it. Of all of the factors considered above, the most central are the meeting locations for each of these connections – field archaeologists met in the informal field; scholars met in strict institutional settings. The relationships that followed reflected those initial meeting places and the subsequent dynamics. Undoubtedly, there were other factors that would dictate the dynamics between Breasted and these men: age, expertise, scholarly goals, and more. But it is important to note that the spaces in which scientific networks begin have a profound effect on the work that each person does and what kind of space they are able to occupy.

The Breasteds arrived home in Chicago in time for Breasted to take up his new position at the new University of Chicago in the autumn of 1895. After the trip, he wrote to his father 'I have acquired the equipment for a great work,' and he went on to prove that over the next four decades (C. Breasted, 1943: 80). He had bought objects for the Museum and his research, but he also gained two important relationships for setting the foundations of his scientific goals. Upon his return to Chicago in 1895, he became Assistant Professor of Semitic Languages and Egyptology, as well as the Assistant Director of the Haskell Oriental Museum. Ultimately, he did much more. His correspondence with other archaeologists, Egyptologists, and scholars grew exponentially from that point on. He heard, largely, from British and German Egyptologists, demonstrating the increasing depth and breadth of his scientific network, thanks to this first and crucial expedition through Egypt for the new PhD. He had built the foundation of his scientific network from the ground up, so for the next forty years he was able to do the work that he had set himself during his first trip to Egypt. He had become an Egyptologist.

Note

1 All correspondence between James Breasted (JHB) and others is located at the Oriental Institute Archives, University of Chicago: Breasted Correspondence (used with permission). They are noted in-text. I would like to thank John Larson and Anne S. Flannery, archivists at the Oriental Institute, for their kind help and hospitality. Thanks also go to Julia Roberts who read and commented on this chapter.

10

Frontier gentlemen's club: Felix Kanitz and Balkan archaeology

Vladimir V. Mihajlović

Histories of archaeology show that our disciplinary knowledge has immensely diverse origins, in terms of its interactions not just with other fields of scholarly inquiry, but within the field of archaeology itself. Routes of communication exist outside 'regular' academic channels and have a great influence on the production and transmission of disciplinary knowledge. Knowledge that is now perceived as canonical has often been conceived through contacts made outside institutional circles and their strict rules. Archaeological knowledge, as well as scientific knowledge in general, like any other form of knowledge, is 'a cultural formation, embedded in wider networks of social relations and political power, and shaped by the local environments in which practitioners carry out their tasks' (Livingstone, 2002: 236; on the social nature of knowledge see Latour, 1996, 2005; Law, 1992). The socio-/geopolitical nature of knowledge that David Livingstone writes about can be clearly seen in the life and work of Felix Kanitz (1829–1904), one of the greatest researchers of the Balkans (and their past) in the nineteenth century. Géza Fehér, the author of the first and still the most comprehensive biography of Kanitz (1932), gave him the flattering nickname 'Columbus of the Balkans'. Kanitz was once perceived as the discoverer of the lands south of the Sava-Danube river boundary, and his books are still 'a veritable mine of rich and scholarly information' on the Balkans – and Serbia and Bulgaria in particular – hence, 'no attempt at summarizing this achievement can do it credit' (Todorova, 2009 [1997]: 71). Kanitz's work on the Balkan lands brought him a great deal of recognition: he was decorated by the Austrian emperor and the Serbian king, and named an honorary member of several learned societies, including the Serbian Learned Society, the Serbian Royal Academy and the Royal Saxon Academy. His publications cover numerous fields of academic inquiry: geography, ethnography, demography, linguistics, folklore, art and, of course, archaeology. Kanitz is celebrated as the author of 'some of the most important early works on archaeology in Serbia' (Novaković, 2011: 387). His enquiries were followed by modern researchers. Petar Petrović and Miloje Vasić, who took part in the large-scale rescue excavations conducted in the Iron Gates gorge, write that 'the validity of documents left by Kanitz could be evaluated best *[sic]*' (Petrović and Vasić, 1996: 15). Kanitz collaborated with the leading scholars and stakeholders of his time, and thus he – at the very least – laid the foundations of Serbian archaeology. Moreover, his influence can still be found in the everyday practice of today's archaeologists: for instance, his site reports are usually a starting point for research, and his writings have been used in the construction of contemporary identities (Babić, 2001: 173; 2002; Cvjetićanin, 2011: 151). Having in mind the important role of Felix Kanitz in Serbian archaeology, the aim of this chapter is to shed light on the context of his research in the field. In order to complete this task, I shall use theoretical insights from geography of knowledge (Naylor, 2002, 2005; Livingstone, 2007; Livingstone and Withers, 2011). Contrary to the widespread belief that science is placeless, authors working in this field have shown that, like 'temporality and embodiment', geography is also a *conditio sine qua non* for scientific endeavour of any kind, since 'spaces both enable and constrain discourse', as Livingstone (2003: 7) nicely puts it. The concept of space in this particular case takes us to the topics of inclusion and/or exclusion, validity, veracity, partiality, etc. Accordingly, this chapter questions the role of geography in both the nurturing and the hindering of Kanitz's scientific understanding and activities, as well as the reception of his endeavours. Finally, having in mind the social origin of knowledge in general, special attention in this chapter is given to the network of contacts Kanitz created; that is, the informal group of people who influenced Kanitz's political, cultural and scholarly views, and consequently left a strong mark on Serbian archaeology as well.

Kanitz's discovery of the Balkans

Felix Philipp Kanitz was born on 2 August 1829 to a 'rich and notable' Jewish family in Obuda, now part of Hungary's capital Budapest (Fehér, 1932; Horel, 2011: 16–17; Teichner, 2015: 7). At the age of 14 he started training as an illustrator at the studio of the famous illustrator Vincenz Grimm (1800–72) in Pest. Grimm was a very important figure in Hungarian artistic circles of the time – he was the founder of the

Pest Art Society (*Pesti műegylet*) – and, likewise, close friend to numerous politicians and scholars in the Habsburg Empire (Horel, 2011: 17; Timotijević, 2011: 94). As a result, while in his youth, Kanitz was presented to Hungarian higher society. For example, Grimm's circle included the Hungarian palatine, as well as the famous topographer, ethnographer and historian József Vincenz Häufler (1810–52), and the archaeologist Ferenc Kiss (1791–1859), who would later become a professor at Pest University. This stimulating learning environment taught Kanitz a wide spectrum of skills: besides artistic illustration, he learned technical drawing, which was and remains one of the basic tools in the disciplines of archaeology and anthropology. Likewise, during his formative years Kanitz became acquainted with the teachings of J.G. Herder (1744–1803) and his Romantic followers. These ideas would become the theoretical framework of Kanitz's writings (Horel, 2011: 17–18; Timotijević, 2011: 92, 100–101).

In 1847 Kanitz moved to Vienna, where he enrolled in the prestigious Academy of Fine Arts. However, just a year later, he left the Academy, though he stayed in the capital, training in lithography at Eduard Singer's workshop (Horel, 2011: 17; Timotijević, 2011: 92). Towards the end of 1848, Kanitz became a correspondent for the *Illustrirte Zeitung* (*Illustrated Newspaper*) in Leipzig, a job he would keep almost until the end of his life (Babić, 2001: 173; Timotijević, 2011: 92). *Illustrirte Zeitung* was the first German illustrated magazine; when Kanitz became its correspondent, it was one of the most prestigious (as well as expensive) illustrated magazines in the German language (Timotijević, 2011: 93).

Even after leaving the Viennese academy, Kanitz continued to expand his intellectual horizons, broadening his knowledge about art and related topics in Munich, Dresden and Nuremberg over the next few years (Timotijević, 2011: 93). Finally, in 1856, after several years of extensive travel throughout Europe (Germany, France, Belgium and Italy), he settled in Central Europe's unofficial centre – Vienna. Symbolically, Vienna was also considered to be at the edge of the Balkans: 'Asien beginnt auf der Landstraße' said Prince Metternich (1773–1859), the German-born Chancellor of the Austrian Empire, in 1820 (Davies, 1996: 55; Sowards, 2004: 42). That is to say, according to the chancellor, Asia, including the Balkan Peninsula, begins on the Landstrasse, the road which leads south and east from the city of Vienna.

His career as a newspaper illustrator brought Kanitz to south-east Europe for the first time in 1858, so he could report on the political upheaval in 'European' Turkey. The uprisings of the Balkan Christians that had started at the beginning of the 1850s in Montenegro and Herzegovina peaked at the end of 1857 and, the next year, spread to parts of nearby Bosnia. The possibility that several peasant rebellions might turn into a fight for national liberation sparked the interest of Europe's Great Powers, as well as its general public (Stolberg, 2008: 68; Todorova, 2009 [1997]: 62). Over the course of his first Balkan excursion, Kanitz visited Montenegro, Bosnia, Herzegovina and Dalmatia. After that trip, he became more and more interested in the nuances of the emerging Eastern Question, until he became completely devoted to the study of south-east Europe, and Serbia and Bulgaria in particular. In short, Kanitz went to the Balkans as a newspaper correspondent but returned to his home in Vienna determined to dedicate himself to a more extensive study of the region.

One year after visiting the western Balkans, Kanitz visited centrally positioned Serbia for the first time; he returned to the country in 1860 and 1861. His first papers on Serbian themes were published in the *Illustrirte Zeitung*, and Kanitz's first major publication – *Die römischen Funde in Serbien* (*The Roman Finds in Serbia*) – was printed in Vienna in 1861 under the aegis of the Austrian Academy of Sciences. The following year he published another, *Serbiens byzantinische Monumente* (*The Byzantine Monuments of Serbia*). Kanitz recapitulated his decade-long impressions and inquiries of Serbia when he published his 1868 scholarly travelogue, *Serbien. Historisch-etnographische Reisestudien* (*Serbia. Ethnographic and Historical Travel Studies*). While *Die römischen Funde* was published under the auspices of the Austrian Academy of Sciences, this last work was supported by the Serbian government to the extent of 300 ducats (Timotijević, 2011: 99, footnote 14). That is to say, in less than a decade Kanitz positioned himself as the crucial author in the field of Balkan studies – he managed to interest both Serbian and Austro-Hungarian governments and their institutions in his work.

Kanitz dedicated the next decade and a half to the study of Bulgarian lands and population. He returned to Serbia again in three consecutive years – 1887, 1888 and 1889 – to continue his previous studies. During these visits, Kanitz also noted the great changes that had occurred in almost every aspect of Serbian society. The country was quickly going through modernisation processes: organisation of public institutions on European lines, urbanisation, industrialisation, railway building, even a change of fashion on the streets of Serbian towns where western hats replaced Turkish fezzes. Over the following fifteen years, that is, until the very end of his life in 1904, Kanitz turned his research into several publications. Firstly, he published a comprehensive work about the Roman heritage of Serbia titled *Römische Studien in Serbien* (*Roman Studies in Serbia* [1892]). This book offers evidence of the magnitude of the changes in Roman archaeology in Serbia. In his first book on the topic, published shortly after his first visit to the country (in 1861),

Kanitz had mentioned around 40 sites, while his later study contains around 300 more. Clearly, though, Kanitz was not solely responsible for this increase/growth, but his discoveries, and the enthusiasm for research that he spread to others, were unquestionably an integral part of the process.

Kanitz's last and the most extensive book on Serbia was published in 1904, in Leipzig – where his first papers on Serbia had been printed – just a few months after his death. *Das Königreich Serbien und das Serbenvolk von der Römerzeit bis zur Gegenwart* (*The Kingdom of Serbia and the Serbian People from Roman Times until the Present*) is the pinnacle of Kanitz's studies of Serbia and covers a time span of more than thirty years. It is also a kind of memoir since it contains, along with his scholarly observations, numerous personal or even intimate moments. Thus, this publication is a collection of diverse data on Serbia as well as a description of the way Kanitz obtained those data.

Finally, it should also be said that throughout this time Kanitz was highly valued as an illustrator – his visual works even found their way into the publications of other authors. For instance, his illustrations adorn Georgina Mackenzie (1833–74) and Adelina Irby's (1831–1911) popular and influential book *Travels in the Slavonic Province of Turkey-in-Europe* (1877), as well as Auguste Viquesnel's (1803–67) *Voyage dans la Turkquie d'Europe* (1868); Kanitz's work can also be found in illustrated magazines of the time, including several articles published by A. Leist in the *Globus – illustrirte Zeitschrift für Länder und Völkerkunde* (Timotijević, 2011: 98).

Nineteenth-century Serbia: between Orient and Occident

Between Kanitz's first visit to Serbia in 1859 and his last, which occurred in 1889, Serbia underwent large-scale political changes. Kanitz's first visit was to a principality officially under the suzerainty of the Ottoman Empire, and his last was to an internationally recognised kingdom. That is, when Kanitz arrived in Serbia for the first time, Serbia was a semi-independent principality but still a part of the Ottoman Empire. By 1867, the country had become *de facto* independent, though formal recognition had to wait until the Congress of Berlin, in 1878. Lastly, in 1882 Serbia became a kingdom. These changes in the formal status of the country were followed by a complete transformation upon a Western and Central European model: the abolition of feudalism, adoption of several constitutions, construction of roads and railways, reorganisation of administration and so forth (Petrovich, 1976; Pavlowitch, 1999; Luković, 2011). First and foremost, the winds of change blew from Serbia's northern neighbour – the Austrian, that is – from 1867 – Austro-Hungarian Empire.

At the same time, the ruling circles of the great Habsburg Empire were also dealing with the Eastern Question posed by the 'sick man of Europe': the Sublime Porte's problems maintaining political control over the Balkans (Anderson, 1966; Bridge, 2002; Sowards, 2004, 209–29). From the end of the eighteenth century onwards, Europe's Great Powers (Britain, Prussia, Russia and Austria) were trying to solve numerous issues connected with the political and economic instability in the Ottoman Empire, in order to maintain the fragile balance-of-power system on the continent. Being the Ottoman Empire's closest neighbour among the Great Powers, Austria (from 1867 Austria-Hungary) was particularly interested in the possibility of seizing power over the Balkan lands hitherto under Ottoman control. Thus, the foreign politics of Austria-Hungary in this part of Europe could be labelled 'frontier colonialism'. The Dual Monarchy's colonial efforts were directed towards its own frontiers, as in the case of the occupation and subsequent annexation of Bosnia and Herzegovina (Donia, 2008; Ruthner, 2008). These lands were surrounded by Austro-Hungarian territory on two out of three sides, in contrast with the more prominent colonial experiences of the British or Spanish Empires. The term 'frontier' here also stresses the fact that Austro-Hungarian colonialism was not, so to speak, colonialism in its full right. However, this colonialism was not particularly unique. Frontier colonialism is just one form of the 'informal imperialism' or 'informal colonialism' practised around the world during the second half of the nineteenth century. Thus, Serbia could be located among those countries in which 'there was an acknowledgement of a need for modernization following western-dominated models', so 'they all had the European presence in their lands' and some of these 'Europeans were trusted to provide advice on political and cultural matters, or even were appointed to Westernise their countries' (Diáz-Andreu, 2007: 99–100).

To sum up, during the second half of the nineteenth century Serbia was politically independent, first as a principality, then, from the 1878 Congress of Berlin, as a kingdom, but under the political, economic and cultural influence of its powerful northern neighbour. Furthermore, all of this accords with the perception of the Balkans as somewhere 'in the middle' that Maria Todorova, a Bulgarian-born US historian, has termed 'Balkanism'. The central idea of her book *Imagining the Balkans* (2009 [1997]) is 'that there is a discourse… that creates a stereotype of the Balkans, and politics is significantly and organically intertwined with this discourse' (Todorova, quoted in Halpern, 2014: 15). The status of the Balkans, according to Todorova, is 'semicolonial, quasi-colonial, but clearly not purely colonial', and thus, their liminal character – which 'invokes labels such as semideveloped, semicivilized, semioriental' (Todorova, 2009 [1997]: 16) – 'could have made them simply an incomplete other; instead they are constructed not as the other but as an incomplete self' (Todorova, 2009 [1997]: 18). The Balkans were/are not perceived as the Other (like the Orient in Edward Said's *Orientalism* (1978)), or even as an incomplete other, but as an 'insufficiently European Europe' (Goldsworthy, 2006: 32); albeit peripheral, the Balkans were still thought of as within European space.

Having in mind the Austro-Hungarian semi-colonial relationship with Serbia it could be said that Kanitz's position was also somewhat liminal. He was an intermediary between the two states, i.e. between two distinct political, economic and socio-cultural entities. Kanitz was a protégé of both the Serbian and Austro-Hungarian governments; his endeavours were in line with the foreign policies of both countries. More specifically, Kanitz received financial support for publishing his works from the small, newly established Principality and the great, old Empire. His information was important to the Habsburg court for its planned expansion *nach Osten* – first economic and cultural and, then, potentially, military. At the same time, Serbia saw a chance to promote itself through Kanitz's writings. The Serbian authorities accepted, helped and honoured Kanitz, and his works were considered one of the cornerstones of the country's representation abroad (Vasić, 1929: 594; Cvjetićanin, 2011: 147; Timotijević, 2011: 108). In fact, Kanitz's publications were an element of Serbia's foreign policy in the second half of the nineteenth century (Teichner, 2015: 11–12).

Kanitz's specific position could be the reason why he, unlike many other observers of those times, restrains himself from either lobbying for or demonising Balkan populations – even though his writings do contain traces of what Todorova elegantly calls 'the specific admixture of nineteenth century romanticism and *Realpolitik*' (2009 [1997]: 62). Despite the fact that Kanitz refused to take sides, his insights and actions, network of contacts and finally the knowledge he produced as well as the reception of that knowledge were all influenced by 'geography', or, more precisely, the geopolitical situation in which he found himself.

Felix Kanitz's Balkan network

The roots of Kanitz's world view, and hence the foundations of the knowledge he created, could be found in the period when he was still learning the craft of engraving and illustration in Vincenz Grimm's studio in Pest. From then on, Kanitz continued to constantly broaden it in the following decades. Before going into details about Felix Kanitz's informal network, his specific position 'in-between' should be stressed once again. Kanitz's liminality cannot be reduced to the realm(s) of foreign policy; it was also social (he was, at the same time, an insider and an outsider in both Austro-Hungarian and Serbian society) as well as disciplinary (his research into the Balkan past could be placed somewhere between amateur antiquarianism and disciplinary archaeology). Therefore, in discussing Felix Kanitz and his network we cannot talk about a 'thought collective' in the narrowest sense: more likely, his position could be placed between those theoretical esoteric and exoteric circles (Fleck, 1979). For instance, while in Nuremberg, Kanitz had the chance to meet Carl Alexander von Heildeloff (1789–1865), a professor of architecture at the Polytechnic School and the City Architect, who specialised in the restoration of medieval buildings. Heideloff's work inspired Kanitz, paving the way for his studies of medieval art in the Balkans, crowned by his publication on Byzantine monuments (Kanitz, 1862). Likewise, after Kanitz settled in Vienna, he and the archaeologist Francesco Carrara (1812–54) became friends. Their friendship immensely deepened Kanitz's knowledge of archaeology (Timotijević, 2011: 93). The roots of Kanitz's interest in Roman archaeology could also be found in the general *Zeitgeist*. During the nineteenth century, states presented themselves as inheritors of ancient Greece and Rome using the symbolic capital (*sensu* Bourdieu, 1984) of the past to expand their cultural and political influence in the present (Diáz-Andreu, 2007: 101). In this particular case, the ruling elites of the Habsburg Empire sought to use the prestige of the ancient world in order to expand the Empire's cultural and political influence in the Balkans. At the same time, Roman heritage served as a proof of Serbia's European-ness (Babić, 2001: 176).

Kanitz's last book on Serbia, *Das Königreich Serbien*, offers crucial insight into this elaborate network of contacts. This three-volume work, as indicated above, essentially contains *Erinnerungen*, or memoirs, of Kanitz's Serbian years. Unlike his previous books, which are more or less scholarly in their essence, *Das Königreich Serbien* is a travelogue, and in accordance with the rules of the genre its narrative is unbounded, sometimes even intimate. This gives us an insight into Kanitz's network, his personal relations with the people who helped him during the decades he spent in the Balkans.

The list includes people whom Kanitz met in Vienna while still preparing for his journeys, as well as those he met in the course of his travels. Ami Boué (1794–1881), a Hamburg-born geologist of French descent, has a prominent place in the first group. Boué is famous for his ground-breaking study *La Turquie d'Europe* (*European Turkey* [1840]), published in Paris just a year before its author moved to Vienna. *La Turquie d'Europe* covers the geography, geology and natural history of the Balkan Peninsula and, hence, it is no great surprise that, among Boué's numerous works, Kanitz found this particular one the most useful (von Hauer, 1882; Kostić, 2011: 6). Kanitz's Viennese circle also included Guillaume Lejean (1828–71), another pioneer researcher of the Balkan lands (Lory, 2011), as well as Heinrich Kiepert (1818–99), one of the most prominent historical cartographers of his time. Kiepert and Kanitz's voluminous correspondence testifies to their contributions to one another's work (Timotijević, 2011: 95).

Finally, Vuk Stefanović Karadžić (1787–1864), one of Vienna's most prominent South Slavs, had a major role in Kanitz's preparations for his Balkan travels. This philologist and linguist, a major reformer of the Serbian language, was highly influential in his homeland, as well as in the German-speaking lands (Duncan, 1970). Karadžić recommended Kanitz to the political and cultural elites of Serbia. He also gave him some highly practical advice: for example, that Kanitz should not reveal his Jewish descent when talking to locals (whether Christian or Muslim) (Lory, 2011: 70). Accordingly, before heading to Serbia Kanitz got in touch with some of the greatest authorities in the emerging academic field of Balkan studies. Contacts that started as Kanitz searched for help and advice in the comfort of Viennese salons sometimes turned into sincere and lasting friendships – especially in the cases of Boué and Kiepert – and more or less frequent written correspondence.

When he finally arrived in Serbia for the first time Kanitz had already been introduced to higher society. Furthermore, Kanitz's interests coincided with those of the elite at that particular moment, as can be seen from an 'announcement' issued by the Serbian Ministry of Education: 'Показатељ овог г. Каниц, молер, путује по Србији да снима и молује знатније манастире, развалине старе, пределе итд., у намери да све то после у збирке изда на свет које ће наше отечество изближе упознати са изображеним народима.' ('The bearer of this [announcement], Mr Kanitz, a painter, is travelling through Serbia to draw and paint notable monasteries, ancient ruins, landscapes, and further on, in order to compile everything and publish it, which shall shortly introduce our homeland to enlightened nations'; my translation) (cited after Kostić, 2011: 3). So that Kanitz might easily finish his task, the Ministry of Education adds that government officials, priests, teachers and 'everyone else' should be at his disposal (cited after Kostić, 2011: 3). As a result, his hosts and guides through bureaucratic labyrinths, as well as through the landscape of Serbia, were numerous government officials: from ministers and mayors to engineers, physicians and priests – everyone indeed.

Of particular importance for Kanitz's archaeological work were people like Janko Šafarik (1814–76) or Jovan Gavrilović (1796–1877), both ethnic Slavs born in Habsburg dominions, who were invited to the newly liberated Principality of Serbia in order to establish state institutions upon Central European models. Šafarik, an ethnic Slovak born in the Hungarian town of Kiskőrös, was educated in Bratislava, Vienna and Prague before coming to Belgrade, where he took up the post of Professor of Physics at the Lyceum of the Principality of Serbia. Šafarik undertook the first archaeological survey in Serbia – in 1868, he went on an archaeological journey to western Serbia, where he conducted small-scale excavations (Milinković, 1998: 427). In 1861 Šafarik left his post at the Lyceum, becoming the director of the National Museum and, then, of the National Library as well (Nikolić, 1979; Milinković, 1985; Novaković, 2011: 387). Together with Jovan Sterija Popović (1806–56), who was also born in the Austrian Empire, Šafarik is the person most 'responsible for the first legal acts to protect the historical heritage' (Babić, 2001: 171). Gavrilović, on the other hand, was a Serb born in the Croatian town of Vukovar (then also in the Habsburg Empire). He took up several important political positions in the establishment of the Principality of Serbia – minister of finance, member of the council of regency after the assassination of Prince Mihailo, member of the State Council, head of the Prince's Chancery. Gavrilović was also the chair of the Serbian Learned Society and had a lively interest in Serbian history (Nikić et al., 2007). Kanitz refers to Šafarik and Gavrilović as friends with whom he shared the same scientific interests as well as the (German) language and social norms. These two very important persons in nineteenth-century Serbian politics, just like the members of his 'Viennese circle', were Kanitz's 'gatekeepers' in a literal as well as metaphorical sense: opening doors for him, both of particular official institutions and of Serbian society in general.

Beside those who, to a degree, were professionally connected to archaeology, many 'laymen' were also of great help to Kanitz's work. Especially important were the county engineers who took Kanitz to archaeological sites and helped him with drawing and mapping (Kostić, 2011: 7–8). Kanitz often praises the intelligence and hospitality of the city and county engineers and does not forget to mention their names: Hesse, Sandtner, Zermann, Valenta, Novak, Riener, Deuster and so forth. In fact, Kanitz's collaborators in Serbia perceived themselves as a distinct group; they thought themselves to be enlightened agents of 'culture' and 'civilisation'. Like Šafarik, Gavrilović, and the engineers mentioned, they were either economic migrants from the Austro-Hungarian Empire or Serbs educated in the Central and Western European universities. Thus, Kanitz's network in Serbia was practically a gentlemen's club. It consisted of people who shared the same language – German – and the same cultural values – from etiquette and customs to fashion and cuisine. In Kanitz's own words (1868: 88): 'Mann und Frau, wie beinahe alle

Ingenieurfamilien Serbiens, eingewanderte Oesterreicher, suchten mit Eifer mich die lange Trennung von deutscher Art und Sitte weniger empfinden zu lassen' ('man and woman, almost everyone from the families of the Serbian or Austrian engineers sought zealously to make me feel as little as possible the effects of my long separation from German manners and customs'; my translation). Moreover, they were also perceived as a distinct group by the natives of Serbia. Kanitz mentions several times in his publications that these engineers were generally called 'Swabians' by the local population, no matter what their ethnic identity really was (e.g. Kanitz, 1868: 268 and 1904: 280).

To summarise, as a focal point of this informal group Kanitz was an intermediary (or when in Central Europe the intermediary) who was presenting the newly resurrected Serbian state to a wider European audience. He was a semi-colonial 'discoverer' of those, to some extent geographically, but even more culturally, distant lands. This 'Columbus of the Balkans' was the provider of new and hitherto unknown information to European scholars; recall Kiepert's work on ancient cartography mentioned above (Timotijević, 2011: 97).

Accordingly, the reception of Kanitz's work also had a double nature. During his lifetime, Kanitz's work was not promoted, translated or printed in Serbia, even though it was financially supported from public funds. Furthermore, his archaeological insights were often ignored or even criticised by the first generation of Serbian archaeologists, such as Mihailo Valtrović (1839–1915) and Miloje M. Vasić (1869–1956) (Mihajlović, 2016: 128–32; Kostić, 2011: 12–13, with references). For instance, the first translation of Kanitz's works in Serbia came only in the 1980s, after his data proved useful during the large-scale excavations in the Iron Gates gorge. Conversely, Kanitz's works were immensely popular in Western Europe and had near-canonical status in Central Europe. They were the first works consulted by basically anyone interested in Serbia, Bulgaria or their respective pasts. Again, as with everything connected with Kanitz, there is also a middle ground in the reception of his work – the Serbs living in the Dual Monarchy. Kanitz's writings were translated into Serbian before the 1980s, but not in Serbia itself – rather, in the parts of Austria-Hungary inhabited by Serbs, that is, more or less, the present-day Vojvodina region.

Conclusion

Kanitz's travels between Vienna and Belgrade were at the same time metaphorical journeys between Austro(-Hungarian) *Realpolitik* and his personal, Romantic ideas about Serbia. The position of Felix Kanitz 'in between' is mirrored in his notion of Serbia. He did not hide his excitement over 'günstigen Umschwung... in opferfreudigster Förderung des Heer- und Schulwesens, im Fortschritt von Wissenschaft, Kunst, Industrie, Land- und Bergbau' (Kanitz, 1904: xii), 'an astonishing vigour... in the development of science, arts, industry, agriculture and mining' (Babić, 2001: 175) among 'das Volk blieb im Kern gesund' (Kanitz, 1904: xii), 'the people [who] remained sane in their core' (Babić, 2001: 175). However, from time to time, Kanitz reminds his readers that this is not-quite-Europe. For example, lamenting the destruction of archaeological sites by looting locals, he notes that 'denn nirgends steht das *Schätzesuchten*, mit und ohne *Zauberformeln*, so stark im Schwunge, wie in Serbien' (Kanitz, 1904: 156) ('nowhere did the quest for treasure, with or without the aid of magical formulas, gain such momentum as in Serbia'; my translation and emphasis). Hence, in spite of his overall sympathies for the country and its people, Kanitz's works on the Roman past picture Serbia 'in terms of a pleasant semi-exotic landscape over which the Roman past is scattered' (Babić, 2001: 175).

The 'semi-exotic' landscape of Serbia is the key to understanding Felix Kanitz's informal network. During the second half of the nineteenth century there was a great demand for knowledge about the Balkans among the newly established Balkan states as well as the old European powers. In this particular case, Serbia searched for a means to present itself to the European audience, in order to eventually take its place in that (imaginary) community of nations. On the other hand, to pursue its semi-colonial politics the Habsburg Empire needed information about the Balkans for its political, economic and (potentially) military aspirations. Meanwhile, both countries searched for relevant, reliable and systematised knowledge – and Kanitz saw his unique opportunity. However, he could not complete this complex and voluminous task on his own, so he was spurred to create a ramified *cum* informal network. This network enabled him to feel at home in these – as it seemed to his Western contemporaries – distant and unknown lands, and members of diasporas – in a way travellers themselves – had an instrumental role in it; people like Vuk Karadžić, one of the champions of the Serbian community in Vienna, on one hand, and Janko Šafarik and county engineers – emigrants from Habsburg dominions – on the other, enabled Kanitz to be simultaneously an insider and an outsider in both Serbian and Austro-Hungarian society.

Helped by the members of the informal network he created, Kanitz reported on the changes that transformed 'Turkey in Europe' to 'European Serbia'. However, he was not just reporting socio-political changes in Serbia – rather, Kanitz took an active role in those changes, especially in the field of Roman archaeology. Faced with the task of creating a knowledge that is simultaneously universal and provincial, it is no great surprise that Kanitz dedicated most of his time to Roman *limes* – a topic which is universal yet provincial in its essence. For that matter, he presented himself as an 'enlightened' and 'civilised' foreigner who was there to patronise the locals. In summary, Kanitz tried to dispel 'magical formulas', consequently transforming the 'quest for treasure' into an academic discipline. This patronising attitude is the reason why he was criticised. He was perceived to be a biased outsider – both ethnically and in a professional/disciplinary sense – while his works were usually neglected by the first generation of Serbian archaeologists.

Nevertheless, after they proved useful during the 1960s, Kanitz's publications have been consulted in excavations of Roman sites ever since. At the same time, they have been used for constructions of contemporary identities. Present-day Serbian archaeologists often refer to Kanitz in their constant effort to prove that Serbia has a claim to European heritage that is, if not greater than that of Western Europe – as signified by the current, state-supported project purporting to show Serbia to be the 'homeland of Roman emperors' (Kuzmanović and Mihajlović, 2015) – certainly equal to it. The irony lies in the fact that in order to cast away the (semi-)colonial image of the state, academic authorities in Serbia turned to Kanitz's publications, which are semi-colonial in their essence (Babić, 2001 and 2002).

Kanitz's travels to and from the Balkans put him in numerous liminal positions: he stood on the narrow line that, at the same time, connected and divided Balkan and 'European' realms. He also stood on the border of the discipline of archaeology itself: an amateur in the field, who created an elaborate Europe-wide network that produced and, following that, transmitted knowledge about the Roman past of Serbia. Through their work, Kanitz and his collaborators tucked their own liminality deep into the fold of Serbian archaeology. Thus, besides being the 'veritable mine of rich and scholarly information' the validity of which is beyond question, the work of Felix Kanitz has brought much more to Serbian archaeology: its theoretical and epistemological foundations.[1]

Note

1 In the course of writing this chapter Thea De Armond, Staša Babić and Monika Milosavljević provided valuable comments and suggestions. V.V. Mihajlović is participating in project no. 177006 of the Ministry of Education, Science and Technological Development, Republic of Serbia.

11

Re-examining the contribution of Dr Robert Toope to knowledge in later seventeenth-century Britain: was he more than just 'Dr Took'?

Jonathan R. Trigg

> Had made dead skulls for coin the chymist's share,
> The female corpse the surgeos purchas'd ware...[1]

This chapter presents a reflection on, and assessment of, the life, career and work of the little-studied seventeenth-century physician and 'Renaissance man' Robert Toope. He is currently, perhaps, chiefly known for his correspondence on wide-ranging, eclectic, subjects with the likes of fellow antiquarian John Aubrey,[2] in addition to Robert Boyle. He also fell victim to later less-than-complimentary references by William Stukeley, who observed in a denigratory fashion that the people local to Avebury referred to Toope as 'Dr Took' (see below). Aubrey, meanwhile, was known to have spent time transcribing the spider-like, looped handwriting of Toope as part of his *Templa Druidum* (e.g. Burl, 2002: 41–2; 2010: 80).

The evidence suggests that Stukeley's, albeit famous, oft-quoted observations were bafflingly ill-considered. Toope was more than merely a product of his time. He seems to have been clearly someone who was subject to periods of intense activity that had great influence on the work of his contemporaries, as well as those antiquaries and academics that followed, and without which we would have far lesser understanding of the archaeological record of the Wessex region. Yet, unlike many fellow antiquarians, for example, he did not publish his own observations, favouring the communication of such to other contemporary scholars.

There are, it seems to me, three forms of network to which Toope's work contributes, and these might be referred to as his contemporaries and near-contemporaries (seventeenth and eighteenth centuries), after which there seems to be hiatus in the archaeological usage of Toope's work up until the second quarter of the nineteenth century and for the ensuing century. Thereafter, he is next and most recently utilised by post-war academics who realise the value of using such observations to inform archaeological work.

As such, we are reliant, therefore, on the material produced by his network of contemporaries and the interpretations of later antiquaries and scholars to understand the importance of such an individual. The nature of the information provided by Toope comes in the form of reporting what would now be considered archaeological material to the members of his social circle who, like Aubrey, clearly had an archaeological bent. There is also evidence of the fact that Toope was truly a Renaissance man, as can be seen by his further interactions with medical men of the time, such as Robert Boyle, but also in his observations relating to the waters of Bath. As much antiquarian material was self-published, and thus done with a particular viewpoint in mind, it seems that a study of this form is of importance in understanding the activities of this period – one indeed which has seen very little work of any sort, and in which the views of established scholars are perhaps overvalued. Not least in the region in which Toope operated (south-west England), much is made of the contribution of many well known individuals, such as those mentioned above, and the contributions of the lesser known actors are less appreciated.

In this chapter, therefore, I consider such issues, highlighting the paramount importance of going back to the original sources in work of an archaeological and similar nature. It also serves to establish such facts as are known about Toope, correct some misinterpretations and introduce some new information in what is more than merely a nuanced biographical essay. In order to do this, a combination of published sources, some less well known than others, and unpublished documentary sources, including hand-written texts, are utilised in order to build up as full a picture as possible. The recovery of these forms of evidence contributes to the understanding of networks of exchange of knowledge as they relate to prehistoric England.

The production of such biographies can produce either hagiographies or character assassinations – in order to assess the work of Toope, here it is hoped that a balanced account may more properly be given. Why was it that Toope did what he did, how and what were and are the implications of his activity? And, as a result, how does such an understanding tell us about both the archaeology of an area and the history of the way it has been interpreted? What was Toope's contribution to knowledge, and what is his ongoing cultural significance? Such questions are more complex than previously considered, I would suggest. This chapter contends that, whilst Toope's techniques were neither scientific nor appropriate (in the modern sense), he has proved vital, indeed seminal, in the recording, interpretation and understanding of archaeology that might otherwise have been missed, unrecorded or unknown. His observations have influenced the subject from the very first time they were set down. In his communications to others within his network, Toope brought to the attention of past and present scholars the very material that he was simultaneously destroying in his particular search for 'ancient remains', and in so doing Toope strongly influenced past and present ideas about archaeology. This is a significant case study for the examination of the paradox in the way that antiquarian views of the past still have an influence on our developing present views.

The earlier and personal life of Robert Toope

Little is known about the personal life of Dr Toope; next to nothing of it has been set down in published form. Thus, for reasons of putting his life into some form of context, here I would like to set down what I have elucidated from a variety of sources. From his will, the only contemporaneous official document which can be confidently associated with the good doctor (National Archives, Ref PRO Prob 11/417), a number of familial relationships can be identified. These may be summarised thus: he was married twice. His first wife was unnamed in the will, but from that marriage there were three daughters, Frances, Anne and Katherine.[3] His second wife was named Grace, and from this union there were two (unnamed) daughters. He is likely to have died in 1693, probate having been granted on 7 July that year. His place of burial seems to have been lost to the ravages of time.

A detailed search of the available records relating to births, marriages and deaths (www.familysearch.org) would seem to indicate that, at the period in question, there were Toope families in Dorset and Wiltshire (East Knoyle and Stourton). For geographical reasons, the Wiltshire branches would likely be more appropriate. Furthermore, assuming that the two Wiltshire Toope branches are part of the same family, there would seem to be a chronological shift from East Knoyle (where documentary evidence exists exclusively for the period 1613 to 28 May 1629) to Stourton (where the same forms of evidence can be found from 26 May 1639 to 22 September 1686 exclusively – see also Bardsley, 1996 [1901]: 758). Significantly (in relation to the preceding paragraph), the name Katherine features as a family name, being the wife of one generation of Robert Toope and the daughter (christened 15 March 1684, baptized 22 March 1686) of the later. From these records, there is the possibility, then, that Dr Robert Toope was born to Robert and Katherine Toope, and baptised on 19 January 1650 (Ellis, ed., 1887: 11).

In later life, and certainly in the period during which he was archaeologically active (at least 1678 to 1685 and down to his death), the unpublished and published record would suggest that Toope was rather a peripatetic individual. It is certain from his will (National Archives, PRO Prob 11/417) that he was recorded as being 'of the city of Bath', but there were many other locations in which he seems to have had periods of residence, and in which he was likely to have made contacts, both people with whom he gained contact and from whom he gained information,[4] and areas on which he was able to report.

On 16 July 1683 Toope acquired Acton Court (the manor house of Iron Acton, Gloucestershire), together with the associated former park, demesne, manorial rights and two fairs:

> Indenture 16 July 1683. Assignment from Mr Oldfield and Mr Atwells to Mr Ridley and Mr Coules in trust for Dr Toope of the House, Park and Demesne of lron Acton. All that Capitall Messuage or Mancion house and scite of the Mannor of Iron Acton with all the Outhouses Courts Yards backsides gardens and Orchards thereto belonging And also the Conygree or Conywarren thereto near adjoyning and all ways waters and casements thereto apperteyning All which said premisses arc scituated together in Iron Acton on the same side of the waye whereon the said Capital Messuage is standing And also all that Parke or ground in Iron Acton to the said Capital Messuage near adjoyning and inclosed with a stone wall and paled commonly called Acton Parke and formerly used as a Parke containing One hundred fifty three acres together with the wood or grove of timber within the same park and all those closes formerly parcell of the said parke one part thereof is now called the Rayles and containe thirty acres the other commonly called the New Grounds and containe twenty eight acres and also all those seven Closes commonly c.alled Brookmeadows and Horsecroft lying near together and adjoyning Acton Parke which arc parceJI of the Demesne lands of the said Mannor and all those two Closes one whereof Iyes on the North side of the parke and containeth five acres and the other lyeth on the Eastward side of the said parke and containeth two acres parcell of a Tenement formerly in the tenure of Edward Short and all those Messuages scituate in Iron Acton Acton llgar Frampton Cotterell and Lattcridge now or late in the severall tenures of Edward Short Samuel Hellier William Walter Thomas Hobbes and Alice Legg widow and all those two fairs holden yearly (and all profits and perquisits of the manor) (WRO 947E/137b)

There were also twelve acres of woodland called Chowle Grove at Frampton Cotterell (WRO 947/1098). Toope did not stay there long, however, selling the property after less than a year, in June 1684, to the barrister, William Gray of Gray's Inn (Manco, 2004: 32).

Furthermore, we can be sure from a letter that he wrote to John Aubrey dated 1 December 1685, that he was at that point domiciled in Bristol (Fowles, ed., 1980: 52–5; also Colt-Hoare, 1821: 63–4; Burl, 2000: 315; Pollard and Reynolds, 2002: 109; Burl, 2010: 73). An anonymous source (anon, 1819: 329–30) places Toope in the Kennet Valley neighbourhood, while Kains-Jackson (1880: 54) suggests that he was resident in Oxford, but by far the most repeated residence for Toope is suggested as being Marlborough. The origin of this suggestion, or at least the first reference to it that I can find, seems to be the article on Avebury by William Long (1858a: 327). This is based on the letter from *Monumenta Britannica* referred to above but noted as being from Bristol. The reasons for associating Toope with Marlborough (or Oxford for that matter)[5] are unclear but they have often been repeated (Davis and Thurnham, 1865; Boyd Dawkins, 1871: 242; Smith, 1884: 169, 172; Cunnington, 1933: 169; Grinsell, 1936: 151; Piggott, 1958: 236, 1962: 4; Cleal and Montague, 2001: 14; Pollard and Reynolds, 2002: 233; Semple, 2003: 79–80; Perks, 2011; Cunnington, n.d.: 12–13), though they may be explained by the contents of his letter to Robert Boyle written from Bath in 1683 (see below, and Birch, ed., 1772: 658) which indicates that he was certainly living in Marlborough by February 1678.

Bath seems to have become his permanent domicile after having stayed there temporarily; he stated in a letter to Aubrey that 'I lodge at ye One Bell in ye Strand and shall tarry 2 or 3 days in Town and your company will be acceptable too' (Fowles, ed., 1980: 55); this accommodation would no doubt have suited Toope for his travels, being on the main coach route to London and elsewhere. Aubrey Burl (2010: 80) states that he certainly had a medical practice there, and this chapter can confirm that evidence, as well as adding some evidence of his character and household, or at least of his having the stature to attest to the presence of servants, and that he was well thought of by fellow medical practitioners.[6] Whilst it is unclear how Toope and Aubrey became acquainted, it does seem that the latter had an interest in the 'healing powers' of the springs of Bath (Burl, 2010: 104), and this might have been the catalyst for the formulation of this aspect of his knowledge networks. Aubrey certainly communicated with the doctors there (Britton, 1845:16–17), and there is no doubt that Toope might be recorded amongst that number (see Guidott, 1708) and is likely to have recommended the cure to the doctors in the area (see also Guidott, 1708: 16–17).

Toope at the Sanctuary (Overton Hill, Wiltshire)

Perhaps the most significant contribution made by Toope to the formulation of archaeological knowledge is the record he made of the presence of an ancient cemetery in Wiltshire. We know that Toope was at the Sanctuary, Wiltshire (a double-ring stone circle of the Neolithic/ early Bronze Age period) in 1678, when he witnessed the discovery of human bones at the site – 'Dr Toope found these bones Ao Dom 1678' (Aubrey in Fowles, ed., 1980: 52–5) informs us of this, from his letter of 1685 showing that he was there again in that year.[7] We are told that at this later point he was living in Marlborough (Long, 1858a: 327, and repeated in Boyd-Dawkins, 1871: 242; Cunnington, 1933: 169; Piggott, 1962: 4) or at Bath (Burl (2000: 315), although see above for a criticism of this view. From this, we learn that, in between Kennet and Overton Hill, on land belonging to 'one Captayne Walter Grubbe' (quoted in Cunnington, 1933: 169), Toope observed, not far from the road, workmen digging enclosure boundaries for French grass (not searching for stones as suggested by Anon, 1819: 329–30) who informed him that many bones had been exposed, although of what form, the workmen knew not. Toope investigated the bones and found them to be human (Fergusson, 1872: 76). This identification can be considered accurate, given Toope's medical qualifications (Fergusson, 1872: 77; Kains-Jackson, 1880: 54).

The next day Toope returned and made his own excavations to recover 'many bushels' of bones in order to make medicines from them which were used to treat the ailments of his patients (see also Long, 1858a: 327; Davis and Thurnam, 1865; Boyd-Dawkins, 1871: 242–3; Smith, 1884: 169; Cunnington, 1933: 169; Grinsell, 1936: 150; Malone, 1990: 26; Burl, 2000: 311; Burl, 2002: 41–2, 61; Burl, 2010: 80).[8] The use of human remains in medicinal treatments was not uncommon at this time (Sugg, 2008; 2011; 2013). Grinsell (1976: 16) uses Toope's activities here and at West Kennet (see below) to extend the expectation of efficacy to all prehistoric tombs, and why not? Toope probably adapted the process used to produce Dr Goddard's Drops – a multipurpose liquid preparation – although given the age of the bones he was dealing with, quite what remained that was of use must be questioned (Cooper, 2004). It seems to me that, as Toope was dealing with bones of some age, he was bucking the trend of the time – where antique material was used, it seems that ancient Egyptian was considered superior. Perhaps he was following the view that older material was superior, or maybe there was less discomfort from the anonymity afforded by unnamed remains. Toope certainly seems to be the sole general practitioner *recorded* as plundering British archaeological material. Whilst there was considerable use of skulls from the Middle Ages onwards, everyone else seems to have been using more recent burials. Followers of the Swiss physician Paracelsus (1493/4–1541) seem to have favoured those that had been buried for not much longer than a year (Richard Sugg, personal communication, 06/02/12).

Toope recorded the condition of the bones – large but 'almost rotten, but ye teeth extreme & wonderfully white hard and sound (no tobacco taken in those days)' (Fowles, ed., 1980: 52–3; Cunnington, n.d.: 12–13; Burl, 2000: 311). He also recorded that they were about 80 yards (*c.* 73 metres – Cleal and Montague, 2001: 14) from 'a larger spherical foundation whose diameter is 40 yards by wch you know ye circuit within this larger Temple there is another orb whose sphere is 15 yards in diameter' (see also Colt-Hoare, 1821: 75, where the letter seems first to have been published; Higgins, 2007 [1829]: xxvi).[9] The land surrounding this monument, Toope records, was level (also Camden, 1722: 208), and underneath were found the burials.[10] He writes that the burials were so close to one another that each skull touched the next. Importantly, he records that he exposed two or three but just to see the nature of the burials, determining that the feet lay towards the prehistoric monument (Smith, 1884: 172)[11] and that they lay less than a foot (30 cm) below the surface. He also noted that radiating out from this group of burials was a further group, at similar propinquity to the former. His view was that the whole site was full of burials (also Colt-Hoare, 1821: 63–4), thus suggesting that the burials covered an extensive area. A later (undated) note to Aubrey states that the name of the field the workmen were enclosing was Millfield (Burl, 2000: 311) in the parish of Avebury (also Colt-Hoare, 1821: 62).

It seems likely that Toope went back to the Sanctuary burial ground to gain further bones to supply more treatments, for there is a note in Aubrey's papers relating to the letter that contains the above information. This states that the first discovery, as we know, was in 1678 and that Dr Toope 'was lately at the Golgotha [i.e. the Sanctuary burial ground] again to supply a defect of medicine that he hath from thence' (Colt-Hoare, 1821: 64; Davis and Thurnam, 1865). In a note appended to the 1685 letter from Toope to Aubrey, the latter states that 'he was lately at ye Golgotha again', meaning that he was there around that year (also Boyd-Dawkins, 1871: 243; Malone, 1990: 27).

William Stukeley (1743: 33) makes reference to this event, but in the opinion of the author, Toope's observations should be taken with a pinch of salt. Stukeley states that the human bones were found in 'digging a little ditch by the temple'. There is, it seems to me, nothing in the original account of the excavations, either by the workmen or by Toope, that suggests the ditch was little. Further, Stukeley states that the 'little ditch was across some small barrows and where there were no barrows' (Stukeley, 1743: 33). Again, Toope's account makes no reference to barrows or any form of earthwork. Indeed, Toope's account is very clear that the site in question was flat. The Sanctuary is, indeed, in near proximity to a vast monumental complex but, as has been shown, and will be shown below, Toope seems to have considered, and shows, himself to be an empirical scientist given the strictures of his time. I would suggest that it is unlikely, although not impossible, that he would have failed to mention barrows, given he stresses the levelness of the plain in his account. Cunnington (1933: 169) informs us that there existed, east of The Ridgeway and to the south of the London road, slight embanked earthworks which may have represented the remains of the enclosures to which the construction of Toope refers. Regrettably, the present landscape means that any such features have probably since been obliterated.

Finally, Stukeley claims that 'Mr Aubrey says sharp and formed flints were found amongst them, arguments of great antiquity' (Stukeley, 1743: 33). Stukeley does not mention what Aubrey source he is dealing with (presumably the at-that-point-unpublished *Monumenta Britannica* – Bodleian MSS. Top gen. c. 24–5), but Aubrey makes no reference to flints, or indeed any other finds, at this site. There is nothing to suggest that Toope was interested in removing anything more than bones from a site, and it is clear that Aubrey states a lack of desire to excavate; regarding Avebury he is absolutely certain in stating 'His Majesty [Charles II] commanded me to dig at the stones [Avebury]... to try if I could find human bones: but I did not do it' (Fowles, ed., 1980: 34). Such a statement defying the monarch would not seem that of a habitual excavator of sites.

Toope's account of the burials at the Sanctuary is the only evidence we have for them; later archaeological work has proved unable to locate any remnants of the site. His 'many bushels' is often considered to be in large part reason for this, although it should also be noted that, after the stones were removed by a Farmer Green in 1724, the ground was later ploughed by a Farmer Griffin (Anon, 1914: 125). As a result, the account is significant, yet it is perhaps indicative of the nature of prehistoric enquiry in Britain in the eighteenth, nineteenth and early twentieth centuries. Toope's observations were frequently referred to in a descriptive manner (e.g. Stukeley, 1743; Camden, 1722; Anon, 1819; Colt-Hoare, 1821; Long, 1858a, 1858b; Thurnam, 1860; Davis and Thurnam, 1865; Boyd-Dawkins, 1871; Smith, 1884; Cunnington, n.d.). When the 66th Annual Meeting of the Cambrian Archaeological Association (Monday 11 to Saturday 16 August 1913, in association with the 60th meeting of the Wiltshire Archaeological Society) convened in Wiltshire, the Avebury environs were the subject of the first excursion, on Tuesday 12th August 1913. At Avebury, the Rev. E.H. Goddard (in the role of tour guide) referred to Toope's activity at the Sanctuary (Anon, 1914: 125). It even made it into the popular travelogue *Roads to the North* (Brooks, 1928: 140–1), although here (and not for the only time, it seems) Toope's observations at the Sanctuary are equated with Avebury.[12] There were, however, few attempts to *interpret* what Toope had observed.

The first such attempt was that of poet Charles Kains-Jackson (1880: 54).[13] In the absence of excavations at the nearby site of Avebury, he used Toope's discoveries to argue that both monuments were burial monuments (also Fowles and Brukoff, 1980: 19; Pollard and Reynolds, 2002: 109). Burl (2010: 111) argues that the human remains Toope exploited were Neolithic commoners, with the upper echelons of society being buried within the stones themselves. It is unlikely, however, that the burials are of Neolithic or Bronze Age in date; the form of burial simply does not fit with known practices of those times. Later, in his attempt to argue for a continuity of burial in circular enclosures from the Neolithic down to the morphology of churchyard enclosures, Hadrian Allcroft (1920: 281–2) cites Toope's account of the burials to argue that the site was on consecrated ground and that the only reasonable explanation for the vast number of burials was that they were specifically brought there for that purpose. Furthermore, he calculated that, allowing for an entrance/avenue, and assuming that the burials encircled the entire site, the burials would be located in a circular area with a diameter of a minimum of 650 feet (*c.* 200 metres), a circumference of 681 yards (*c.* 623 metres), and which enclosed 7½ acres (*c.* 3 hectares). Allowing for the close setting of the burials that Toope described, Allcroft argued that as many as 2,000 burials were possible.

It seems most likely that what Toope described is an, as yet, undated rural Anglo-Saxon cemetery. There is, for example, a mass burial of the period in a similar landscape context elsewhere on the Ridgeway (Williams, 2015). An early reference which utilises Toope's observations to suggest this period is that of Mr H.C. Brentnall (quoted in Cunnington, 1933: 169),[14] who argued that the burials were possibly the remains of warriors killed in the battle *aet Cynetan* between the Danes and Saxons in 1006 CE (*cf.* Cleal and Montague, 2001: 14; Pollard and Reynolds, 2002: 234). It is Fergusson (1872: 77–8), however, who first makes the argument based on Toope's observations of the quality of preservation of the bones and his record of the forms the burials took. This is interesting and, following on from his and Kains-Jackson's (1880: 54) observations that Toope's anatomical findings could be trusted, it is interesting to note that one of the notes he made was that the bones were large (Fowles, ed., 1980: 52–5). The suggested lack of small (i.e. child) remains adds to the possibility of this being a conflict-related cemetery. There are, however, factors that militate against a battle-related date of 1006 CE for the burials. For example, conflict-related deaths of this period tend to be buried in mass graves, rather than being individual burials in cemeteries in the ordered way Toope suggests. Moreover, the prehistoric barrows, the Sanctuary and other features form part of the boundary between Avebury and West Overton, and Pollard and Reynolds (2002: 234) suggest that this feature was created around 939 CE. Importantly, they go on to argue that the only burials dating to this period related to executed individuals. Whilst it is impossible without the skeletal remains to identify a cause of death, the number of inhumations suggested by Toope's account make this an unlikely scenario. Returning to the issue of the dating of the cemetery, it seems possible that it was early Anglo-Saxon (Pollard and Reynolds, 2002: 233); the lack of grave goods suggested by Toope's account makes it more likely to be late seventh century or later (Pollard and Reynolds, 2002: 234).

The view that it was a Saxon cemetery was followed by Stuart Piggott (1958: 237), while Cleal and Montague (2001:14) suggest the burials were likely to be medieval or Roman,[15] and Semple (2003: 79) suggests the possibility that they were early medieval, and possibly conversion period. At the same time, they argue that the burials were probably extended, rather than contracted, and likely supine (Pollard and Reynolds, 2002: 234). To this author, the fact that there was a definite statement of the orientation 'towards the Temple' (Fowles, ed., 1980: 52–5) confirms this fact quite definitively. The cemetery could, of course, be of any date (Williams, 2015); on the basis of the condition of the teeth referred to by Toope, Burl (2002: 141) argues that they are Neolithic, although I cannot see a reason why they could not be of any date prior to tobacco and the greater use of sugar.

Toope's account has also been used to place the location of this possible cemetery. He noted that the burials were located in flat land and, as Cleal and Montague (2001: 14) note, the most level ground surrounding the Sanctuary is to the north of the monument: on the other sides the ground slopes away. This would make it seem likely, however, that the burials would have lain *over* the road from the monument, as evidenced by the fact that, by the time of the Andrews and Dury map of Wiltshire (1773) the road followed the course of the current A4, and Samuel Pepys observed that the stones of the Sanctuary were visible from the road at the time he visited the area (1668). With this in mind, as Cleal and Montague (2001: 14) observe, it is indeed strange that the normally meticulous Toope did not note the fact that the burials were separated from the monument by the road. Excavations of the area have failed to locate any trace of the cemetery. Regrettably, it seems that this area, if it was the location of the burials, was probably archaeologically obliterated when the Ridgeway Café was built there; had the burials been to the east, they would have fallen victim to quarrying. Thus, any remains would probably have been removed then (Cleal and Montague, 2001: 14; Pollard and Reynolds, 2002: 234). While unlikely to have been completely removed by Toope, he would certainly have greatly lessened the number present (Williams, 2015). It is important, however, to consider the chronological distance between the two periods of Toope's activity at the Sanctuary, and entertain the possibility suggested by Richard Sugg (2011: 92) that he had returned in between these dates to collect further samples, which will have further lessened the chance of finding extant remains.

The landscape context of the remains documented by Toope has been noted by Semple (2003: 79). She observes that the burials are frequently referred to as being related to the Ridgeway, 'a prehistoric route that was of great significance for communication lines in the Anglo-Saxon period' (2003: 79) but also states that the cluster relates to the crossroads between the Ridgeway and the Mildenhall to Sandy Lane/Verlucio road, also in use in the Anglo-Saxon period. Thus, the cemetery 'was sited at a location which commanded views of two major communication routes'. It is perhaps also of note, given the longevity of land divisions, that the location marks the boundary between Avebury and West Overton parishes (Semple, 2003: 88).

Dr Toope and the long barrow at West Kennet

In contrast to the detailed description of the Toope material from the Sanctuary described above is the other archaeological site with which he is principally associated: West Kennet Long Barrow, also in Wiltshire. Perhaps as a result of the excavations of Stuart Piggott (1958; 1962), although more likely the result of a more complete scientific archaeological history, where the site has been excavated recording its biography, Toope may be most associated with this site. This is ironic, for there is no direct historical evidence for his presence here; he does not refer to it in his correspondence, nor is there any direct or contemporary reference to him being at the site (for example by Aubrey, as has been assumed, e.g. Malone, 1990: 26).

However, in Stukeley's unpublished papers (Bodleian, Gough Maps 231) is the note that 'Dr Took, as they call him, has miserably defaced South Long Barrow [West Kennet Long Barrow] by digging half the length of it. It was most neatly smoothed up to a sharp ridge, to throw off the rain, and some of the stones are very large.' Stukeley also recorded the evidence in the form of drawings which he made in 1723–24 (these are reproduced in Piggott, 1962: plate IIa and b). Dr Took is unanimously equated with Toope (e.g. Thurnam, 1860: 408; Piggott, 1958: 236; Malone, 1990: 26). Based on the evidence provided to and by Stukeley, demonstrating that the excavations were prior to 1723, Toope does indeed seem the most likely culprit (*cf.* Piggott, 1962: 4).[16]

When Piggott (1958; 1962) excavated the barrow in 1955–56, he sectioned the mound, revealing the considerable extent of Toope's diggings. Virtually all of the south side of the barrow had been targeted, as was revealed by the presence of craters, some of which had been backfilled with prehistoric and Roman material, together with clay pipe fragments, and the disturbance was clearly visible in section (Piggott, 1962: 4). Considerable attention had been paid to the area of the forecourt and passage of the tomb. Indeed, the capstone of the south-east chamber had been removed and dragged off to the south to allow attempted access to the contents (Piggott, 1958: 236; 1962: 4). The north-east chamber had also been dug into, but in this case, the architecture had been left intact (Piggott, 1962: 4).

It is presumed, quite reasonably (Piggott, 1958: 237; 1962: 4; Malone, 1990: 26) that these excavations were also attempts to retrieve human bones for medicine. This time it seems Toope was unsuccessful, however, since owing to the way the chambers had been backfilled in antiquity, and the formation of the mound, Toope excavated only three feet (c. 91 centimetres) into an eight-feet-deep (c. 2.43 metres) deposit above the primary human remains. However, this suggestion is frequently ignored, and it is assumed he did gain burials from this site (e.g. Dyer, 1990: 55; Perks, 2011: 10; also suggested by Burl, 2002: 278). If, of course, burials were removed at this stage, or indeed in later antiquarian (i.e. pre-Thurnam) periods, we do not know the numbers involved.

It has long been considered fact that the first excavation of West Kennet was by John Thurnam in 1860 (Piggott, 1958: 235; 1962: 4); however, I wonder whether Toope's interventions (given the degree of the damage noted, they must have taken place on a number of occasions) might be considered such. In noting the observations made by Toope regarding the orientation and setting of the burials at the Sanctuary, there is the possibility that anything he might have observed at West Kennet might have been similarly recorded, but not have survived. Toope might equally have recorded non-burial-related material. It is interesting to note, however, that there is nothing in Aubrey's papers to record Toope's potential activity at West Kennet. This suggests a number of possibilities – either the damage was not done by Toope, or perhaps Aubrey did not know of the activity (for example, Toope was the perpetrator, but chose not to report it, perhaps because he did not find human remains which was what he considered important). It seems unlikely that any information sent to Aubrey by Toope would not have been retained by someone as reliable as Aubrey and, given that Toope refers to Aubrey as his 'worthy friend' (Fowles, ed., 1980: 52–3), it seems likely that the two would have been in contact if either thought it important.[17]

Dr Toope and Robert Boyle

In considering the intellectual qualities and contemporary standing of Dr Toope, it is beneficial to consider his contributions outside the archaeological sphere. Of note amongst these is his correspondence with the Honourable Robert Boyle (Davis and Thurnam, 1865). Toope wrote to Boyle on at least one occasion – 5 April 1683 (Maddison, 1958: 172 and 191). It is worth recording in some detail here:

> Honoured Sir
> Since my return into the country I have been very ill in a fever, or else this had (to promise) flown sooner into your hands… I was sent for to one Mrs Corle, of Freshford near Bath, who laboured in a fever; and I took in my pocket a whole ounce of Spanish flies pulverised *grosso modo*, for I usually draw blisters with little bags; and after I had filled two or three bags, which could contain no more than a drachm, I lapped all up in a double paper and stuck it between an iron bar and the glass in the gentlewoman's chamber window, the window looking towards the south, and I went off and left my flies as before. About thirteen or fourteen months after, April or May come twelve month, I was sent for to one Mr Sliper of Tunbridge (three miles from Freshford) brother to the gentlewoman Mrs Corle; and when I came I found my patient Mrs Corle there; I told her I must blister her brother, and spoke to have the apothecary sent for, on which she told me I could have brought old ones; if not young ones enough; for, said she, cleaning my chamber window two days since, I took down a paper of Spanish flies you left there after my last sickness; and when I opened your paper, there were multitudes of little flies like your old ones, and being afraid of their motion, I hastily and rudely lapped it up again, and put the paper where I had it. Then I became warm and impatient to see the phoenix from its ashes; she freely offered her man to fetch me the paper, which I accepted of, and then saw of my own eyes, and many others besides myself, thousands of them creeping and crawling about most exactly shaped to the old flies, long and small their wings, as long as their bodies, but of a very faint glittering and shining colour. I kept some of them in boxes with fruits and leaves, and they waxed bigger, and the bigger they grew, various colours came on. My children observing me often visiting and feeding my little fry, in my absence destroyed my stock.

> I pray Sir give me a taste of that blessed oil, of which you promised the way of consecting [?] and that of refining tin, given you by one Wilden as I take it, for I resolve to work upon that body, so much have I seen to encourage me, and when I am at a loss or stand, I shall beg your assistance, and whatever I do in this kind or any other way, I will communicate and return it back again, as the little rivulet pays the main ocean.
>
> Sir, you must pardon the rudeness of my long nonsensical story, but, if I mistake not, such stories as these ought to be told, that no such circumstances be omitted, for many times the whole matter lies couched under circumstances (though the case does not here so appear) you may give it what philosophical dress you please. Worthy and honoured Sir, I am, yours to command, Robert Toope
>
> To me at the post house, Bath, Somersetshire. *Vise remitte vale.*

In addition to this, it is tempting to imagine further contact, for Boyle, a lifelong sufferer from nosebleeds, was a user of skull medicine. When he used the practice, he found that it seemed to work even before being applied in the usual fashion. He had suffered an unexpected, violent nosebleed whilst at the house of his sister. Having obtained from her the 'moss' from a skull he was going to apply it in the usual manner, by insertion up the nose. Before doing so, however, he was tempted to see if merely holding it in his hand would prove efficacious. The nosebleed not only stopped, but he was not troubled by a nosebleed for some years thereafter (Boyle, 1675).

Discussion

Few people are remembered over 300 years after their death. That Toope is remembered is thanks not to his personal papers, or published works. It is thanks to networks of knowledge distribution. In the first case, it was his communications with contemporaneous scholars – the likes of John Aubrey and Robert Boyle – and there is no reason to discount the possibility that there were more contributions than those known, or that Toope had a wider circle of correspondents. Next, we can see the ways in which his work was used by first antiquarians and then more recent scholars, which has served to maintain his presence in scholarship.

The nature of the correspondences that Toope engaged in was the primary form in which scientific research was communicated in the seventeenth century; letters from one scholar to another (Maddison, 1958: 129). It is to be noted, also, that Toope's letter of 1678 (Fowles, ed., 1980: 52–4) does state that this communication is the precursor to the communication of further 'things of this nature', suggesting that there was more communication between the two which has not survived, and it is possible that this might be related to the Sanctuary, West Kennet, or indeed some other site altogether. This fact is important, however, since it does demonstrate that the material communicated was not a one-off.

We can infer also that the information provided by Toope was reliably sourced. The informants used by Aubrey to provide sources of information for his writings were known to have been such, and there is no reason to consider Toope otherwise. That Toope knew in which places to dig in order to retrieve human bones suggests he may have studied the activities of antiquaries either in print or in person. Furthermore, from his examination of the diggings at the Sanctuary, it seems to be the case that he made considered observations, which he reported (*cf.* Fergusson, 1872: 77). The number of times he was cited by later antiquarians and more recent archaeologists, and even more popular sources, is perhaps evidence of the significance of these observations. Stukeley, perhaps, had an axe to grind when scathingly recording him as 'Dr Took'. Toope was, it seems, much more than that, and one can only wonder at what other observations he might have made at other sites, but which have not come down to us in the present day.

Notes

Several people and institutions have contributed to this chapter and the research from which it derives; these include the Society of Antiquaries of London (who provided a generous grant for research in Oxford from the William Lambarde Memorial Fund), the University of Glasgow and the University of Liverpool. Dr Richard Sugg has, on a number of occasions, provided a ready base for discussion of a number of points. Finally, I am grateful to my fellow editors for the invitation to produce this chapter.

1 Richard Savage, 'The Progress of a Divine: a satire', 1735.
2 He was not, however, one of Aubrey's *amici*. Still, Aubrey was well known for corresponding with fellow antiquarians in the production of *Templa Druidum* (Burl, 2010: 182).
3 There is a Katherine Toope, baptised 22 September 1686, daughter of Robert and Hannah Toope, in Stourton (Ellis, ed., 1887: 18).
4 It seems likely to me that someone as mobile as Toope would have relied upon local knowledge to inform him on the nature of local antiquities which he could have used as sources of human remains.
5 Toope signed himself with the suffix MD, although it is not evident where he was educated. A link with Oxford University has been considered, but a search of the directory of alumni for the appropriate period (Foster, 1891) has located only one Robert Toope, of Trinity College; but he matriculated 8 December 1658 and probably died 18 September 1671 (Foster, 1891: 1496).
6 '...among others, I formerly observed a modest servant of Dr Toope, sometimes since a laudable practitioner at the Bath, who knowing her self *[sic]*

honest desir'd me not to report there was any Milk *[sic]* upon the Blood *[sic]* for fear it might be thought that she was with Child *[sic]*' (Guidott, 1708: 55).

7 This activity is frequently (e.g. Sugg, 2013: 829) misinterpreted as 1684 and/or being at West Kennet.

8 Although Grinsell (1936: 150–51) states this was in the early eighteenth century, this must be an error.

9 The normally scrupulous Colt-Hoare (1821: 79) described the Sanctuary erroneously, suggesting that, from this passage in *Britannia,* Camden must have seen Aubrey's *Templa Druidum (Monumenta Britannica)*, and thus Toope's letter. On the basis of the dates (Camden died in 1623), this cannot have been the case; more likely it was inserted by Edmund Gibson in his 1695 translation of the text from Latin into English (*cf.* Fowles, ed., 1980: 23), to which he made considerable additions, which suggests that Gibson may have had access to Aubrey himself (Aubrey dying in 1697). There were further editions of Gibson's version in 1722, 1753 and 1772. Equally, it could have been first reported in Richard Gough's (1789) edition of *Britannia*; versions would have been available to Colt-Hoare. However, in the collections at Stourhead (the ancestral home of Colt-Hoare), in addition to Camden's 1587 edition (the second – National Trust NT 3006341), there is a version dated 1730, and so presumably Gibson's 1722 version (National Trust NT 3196402). The error is repeated by Higgins (2007 [1829]: xxix), who himself probably got it from Colt-Hoare, or a reading thereof.

10 Long (1858b: 66) suggests that Aubrey's dimensions for the stones at the Sanctuary are confirmed by Toope; evidently the opposite must be the case.

11 Here, Smith does not reference the source of his information, and also states that Toope was 'then living at Marlborough'; without a reference it is not clear whether Toope was genuinely living or just staying at Marlborough.

12 It is of note that Brooks makes reference to other antiquarian sources, not least Camden and Leland.

13 It is interesting to note that Kains-Jackson (1880: 54) is the only person to refer to Toope as 'one of the Carolean antiquaries'. Perhaps the absence of such a sobriquet is reasonable, as there is no evidence that Toope had any interest in archaeological sites except for their likelihood of producing human remains. Of note however, in relation to this chapter, are the observations he made as a by-product of his medicinal work.

14 It is noteworthy that this observation is frequently attributed to Maud Cunnington (e.g. Cleal and Montague, 2001: 14; Pollard and Reynolds, 2002: 234). It was not; rather Mr H.C. Brentnall, as she noted (Cunnington, 1933: 169).

15 The possibility of a Romano-British date for the cemetery is also entertained, albeit briefly, by Pollard and Reynolds (2002: 234)

16 The author is taken by the view (Fowler and Harris, 2015) that Toope's excavations represent a new imagining or interpretation of the tomb.

17 Given the thoroughness with which Aubrey surveyed the Avebury region (see Fowles, ed., 1980), it is surprising that there is no mention of West Kennet Long Barrow, either in drawings or in writing.

Bibliography

Archival sources

Academia Belgica, Rome, Cumont–Vermaseren correspondence
Allard Pierson Museum, Amsterdam, Vermaseren archive
Antikensammlung PKB: Furtwängler's gem catalogue manuscript
Antikvarisk-topografiska Arkivet (ATA), Stockholm
  Hanna Rydhs och Bror Schnittgers arkiv, vols 4, 5
Archives of Associazione Internazionale di Archaeologia Classica, Rome
Archives of Charles University, Prague
Archives of Soprintendenza Speciale per i Beni Archeologici di Roma
  Roma. S. Prisca. Lavori di scavo e restauro eseguiti dall'Ist. Storico Olandese (1953–59)
Archivio Centrale dello Stato, Rome
Bayerisches Staatsbibliothek
  Nachlass Furtwängler
  Nachlass Furtwängler-Scheler
Biblioteca Angelica, Fondi Barnabei (Rome)
  Busta 237/3, Montelius (letters from Oscar Montelius to Felice Barnabei)
The Bodleian Library, Oxford, Bodleian MS
  Top gen. c. 24–5, John Aubrey's *Monumenta Britannica*
  Gough Maps 231
Deutsches Archäologisches Institut
  Nachlass Brunn
  Nachlass Furtwängler
French National Archives, Paris
Gothenburg University Library, Gothenburg
  Kvinnohistoriska Samlingarna, Hanna Rydhs Arkiv A 12, I:6, I:15, II:6
Masaryk Institute and Archives of the Academy of Science of the Czech Republic, Prague
The Metropolitan Museum of Art (MMA), New York

Archives – Correspondence; Marquand, Henry Gurdon, 1819–1902
Archives – Office of the Secretary Correspondence files 1870–1950
Archives – Henry Gurdon Marquand Papers
Museum of Far Eastern Antiquities Archive
JA Rydh to Gunnar Andersson, 18 May 1929
Gunnar Andersson to Bernhard Karlgren, 10 March 1931
Museum of Fine Arts (MFA), Boston, Mass.
Department of Art of the Ancient World Archives
The National Archives, Kew
PRO Prob 11/417 Will of Robert Toope
Nationaal Archief, The Hague, archive of NWO/ZWO
Newberry Library, Chicago
Archives – Charles L. Hutchinson Trustees President Correspondence
Riksantikvarieämbetets arkiv, Antikvarisk-topografiska arkivet (ATA), Montelius-Reuterskiölds samling (Stockholm)
F2B Almanackor och dagböcker m.m (calendars and diaries), F2B 1a, 1864–1879
F4 a1 Fotografier (private photos)
Riksantikvarieämbetets arkiv, ATA, Oscar Montelius arkiv (Stockholm)
E1a Svenska brevskrivare (correspondence), E1a, vol. 42, Vi–Vie
E1a Svenska brevskrivare (correspondence), E1a, vol. 4, Bergen–Bl
E1c Utomnordiska brevskrivare (correspondence), E1c, vol. 63, B–Bo
F1b Arbetsmaterial (work material), F1b, vol. 148
The Royal Library, Copenhagen
1929: Letter from Hanna Rydh to Lis Jacobsen, 6 August 1929
Royal Netherlands Institute in Rome (KNIR), Vermaseren archive
The Smithsonian Institution, Washington DC
Collected Letters on Ethnology (CLE), Record Unit 58. Smithsonian Institution Archives
Charles Rau Papers (CRP), Record Unit 7070. Smithsonian Institution Archives
Stadsarchief, Amsterdam, van Lansdorp archive
Uppsala Universitets Bibliotek (UUB)
Collection of Swedish portraits, photo id. 11254
Uppsala University Archives, Uppsala
Letters from Hanna Rydh to Stina Rodenstam, 2 and 19 October 1922
Wiltshire Records Office
WRO 947E/137b, Deeds Relating to a House and Closes (Described) in Iron Acton
WRO 947/1098, Deeds of Twelve Acres of Woodland Known as Chowle Grove at Frampton Cotterell, part of the Manor of Iron Acton

Internet sources

www.familysearch.org
www.scb.se/hitta-statistik/sverige-i-siffror/prisomraknaren/
Bouzek, J. 2012. Jindřich Čadík. Available from http://ukar.ff.cuni.cz/node/163, accessed on 31/03/16.
Cooper, P. 2004. Medicinal properties of body parts. *The Pharmaceutical Journal*, 18 December 2004, available from https://webcache.googleusercontent.com/

search?q=cache:EbnvU2aOYSMJ:https://www.pharmaceutical-journal.com/news-and-analysis/features/medicinal-properties-of-body-parts/20013612.article+&cd=68&hl=en&ct=clnk&gl=uk, accessed on 24/02/11.

Donia, R.J. 2008. The proximate colony. Bosnia-Herzegovina under Austro-Hungarian rule. Available from www.kakanien-revisited.at/beitr/fallstudie/rdonia1.pdf, accessed on 25/03/16.

École française d'Athènes. 2014 [updated 2017]. Membres étrangers de lEFA. Available from www.efa.gr/index.php/fr/membres-scientifiques/anciens-membres-etrangers, accessed on 15/11/17.

Epigraphic Survey. 2014. The Chicago house method. Available from https://oi.uchicago.edu/research/projects/epi/chicago-house-method, accessed on 28/03/18.

French, W.M.R. 1889. *1889 Travel Notebook*. Available at The Art Institute of Chicago, http://aic.onlineculture.co.uk/ttp/, accessed on 10/07/19.

Hansson, U.R. 2014. Adolf Furtwängler (1853–1907): the Linnaeus of Classical Archaeology. *Antiquity* 88(342). Available from Project Gallery, http://journal.antiquity.ac.uk/projgall/hansson342, accessed on 04/07/19.

Jarnicki, P. 2016. On the shoulders of Ludwik Fleck? On the bilingual philosophical legacy of Ludwik Fleck and its Polish, German and English translation. *The Translator* (published online 23 March 2016). Available from www.researchgate.net/profile/Pawel_Jarnicki/publication/299374138, accessed on 04/07/19.

Jonasson, D., D. Amnéus, U. Flock, P. Rosell Steuer and G. Testad. 2004. Bekräftartekniker och motstrategier – sätt att bemöta maktstrukturer och förändra sociala klimat. Available from www.jamstallt.se/docs/ENSU%20bekraftartekniker.pdf, accessed on 04/07/19.

Perks, A.M. 2011. Stonehenge and its people: thoughts from medicine. Available from https://circle.ubc.ca/handle/2429/33585, accessed on 22/02/14.

Ruthner, C. 2008. Habsburg's little Orient. A post/colonial reading of Austrian and German cultural narratives on Bosnia-Herzegovina, 1878–1918. Available from www.kakanien-revisited.at/beitr/fallstudie/cruthner5.pdf, accessed on 25/03/16.

Stevenson, A. 2015. Abu Bagousheh: father of pots. In: A. Stevenson, ed. *The Petrie Museum*. London: UCL Digital Press. Available from https://ucldigitalpress.co.uk/Book/Article/3/20/0/, accessed on 21/12/17.

Printed sources

Abt, J. 2011. *American Egyptologist: the life of James Henry Breasted and the creation of his Oriental Institute*. Chicago: University of Chicago Press.

Adams, J.M. 2013. *The Millionaire and the Mummies: Theodore Davis's Gilded Age in the Valley of the Kings*. New York: St Martin's Press.

Agnew, H.L. 1993. *Origins of the Czech National Renascence*. Pittsburgh, Penn.: University of Pittsburgh Press.

Alberti, S.J.M.M. 2003. *Conversaziones* and the experience of science in Victorian England. *Journal of Victorian Culture* 8(2), pp. 208–30.

Alberti, S.J.M.M. 2007. The museum affect: visiting collections of anatomy and natural history. In: A. Fyfe and B. Lightman, eds. *Science in the Marketplace*. Chicago: University of Chicago Press, pp. 371–403.

Alberti, S.J.M.M. 2009. *Nature and Culture: objects, disciplines and the Manchester Museum*. Manchester: Manchester University Press.

Alberti, S.J.M.M. 2017. *Nature and Culture: objects, disciplines and the Manchester Museum*. 2nd edn. Manchester: Manchester University Press.

Alexander, K. 1994. A history of the ancient art collection at the Art Institute of Chicago. *The Art Institute of Chicago Museum Studies* 20(1), pp. 6–13.

Allcroft, A.H. 1920. The circle and the cross. *The Archaeological Journal* 77, pp. 229–308.

American Philosophical Society. 1799. Circular letter. *Transactions of the American Philosophical Society* IV (1793–8), pp. xxxvii–xxxviii.

Anderson, M.S. 1966. *The Eastern Question, 1774–1923: a study in international relations*. London and New York: Macmillan and St Martins Press.

Andersson, J.G. 1929a. The origin and aims of the Museum of Far Eastern Antiquities. *Bulletin of the Museum of Far Eastern Antiquities* 1, pp. 11–27.

Andersson, J.G. 1929b. On symbolism in the prehistoric painted ceramics of China. *Bulletin of the Museum of Far Eastern Antiquities* 1, pp. 65–9.

Anon. 1819. Review of new publications. *The Gentleman's Magazine and Historical Chronicle*, 89(2), pp. 329–47.

Anon. 1889. Art Institute of Chicago. *Preliminary catalogue of metal work, Graeco-Italian vases and antiquities*. Chicago: Early and Halla Print Co.

Anon. 1903. Half million expected. *Boston Evening Transcript*, 30 January, p. 10.

Anon. 1914. Report of the sixty-seventh annual meeting held at Devizes, Wilts, August 11th to August 16th 1913. In conjunction with the sixtieth meeting of the Wiltshire Archaeological Society. *Archaeologia Cambrensis*, XIV, pp. 113–204.

Appadurai, A. ed. 1986. *The Social Life of Things: Commodities in cultural perspective*. Cambridge: Cambridge University Press.

Armellin, P. and M. Taviani, 2017. *Una rilettura dellarea archeologica presso la Chiesa di Santa Prisca*. In: Alessandra Capodiferro, Lisa Marie Mignone, Paola Quaranta, eds, *Studi e scavi sull'Aventino, 2003–2015*. Rome: Edizioni Quasar, pp. 131–47.

Arnold, B. 1990. The past as propaganda: totalitarian archaeology in Nazi Germany. *Antiquity* 64, pp. 464–78.

Arthurs, J. 2012. *Excavating Modernity. The Roman past in Fascist Italy*. Ithaca, NY: Cornell University Press.

Arthurs, J. 2015. The excavatory invention: archaeology and the chronopolitics of Roman antiquity in Fascist Italy. *Journal of Modern European History*, 13, pp. 44–58.

Arwill-Nordbladh, E. 1987. Oscar Montelius and the liberation of women: an example of archaeology, ideology and the early Swedish Womens Movement. In: T.B. Larsson and H. Lundmark, eds. *Approaches to Swedish Prehistory: a spectrum of perspectives in contemporary research*. Oxford: BAR-IS, 500, pp. 131–42.

Arwill-Nordbladh, E. 1995. Paradoxen Hanna Rydh: arkeologi, emancipation och konstruktion av kvinnlighet. In: J. Nordbladh, ed. *Arkeologiska liv*. Gothenburg: Institutionen för arkeologi, Göteborgs Universitet, pp. 77–103.

Arwill-Nordbladh, E. 1998. Archaeology, gender and emancipation: the paradox of Hanna Rydh. In: M. Diaz-Andreu and M-L. Stig Sørensen, eds. *Excavating*

Women: a history of women in European archaeology. London: Routledge, pp. 166–74.

Arwill-Nordbladh, E. 2005a. Tankar kring en professionalisering: Hanna Rydhs arkeologiskt formativa tid. In: J. Goldhahn, ed. *Från Worm till Welinder: åtta essäer om arkeologins disciplinhistoriska praxis*. GOTARC Serie C. Arkeologiska skrifte, 60. Uddevalla, pp. 109–42.

Arwill-Nordbladh, E. 2005b. Där fädrens kummel stå: om fornminnesplatser som offentligt rum. In: R. Engelmark, T.B. Larsson and L. Rathje, eds. *En lång historia: festskrift till Evert Baudou på 80-årsdagen*. Umeå: Institutionen för arkeologi och samiska studier, pp. 35–49.

Arwill-Nordbladh, E. 2008. Twelve timely tales: on biographies of pioneering archaeologists. *Reviews in Anthropology* 37(2–3), pp. 136–68.

Arwill-Nordbladh, E. 2013. Ethical practice and material ethics: domestic technology and Swedish modernity exemplified from the life of Hanna Rydh. In: S.M. Spencer-Wood, ed. *Historical and Archaeological Perspectives on Gender Transformation: from private to public*. New York: Springer Press, pp. 275–303.

Ås, Berit. 1978. Hersketeknikker. *Kjerringråd* 1978(3), pp. 17–21.

Avenarius, A. et al. 1992. Byzantská studia v Československu. In: *Dějiny Byzance*. Prague: Academia, pp. 472–5.

Babelon, E. 1900. Review of Furtwängler 1900. *Journal des Savants*, pp. 445–7, 594–609, 652–68.

Babić, S. 2001. Janus on the bridge: a Balkan attitude towards ancient Rome. In: R. Hingley, ed. *Images of Rome: perceptions of ancient Rome in Europe and the United States in the modern age*. Portsmouth, RI: Journal of Roman Archaeology, pp. 167–82.

Babić, S. 2002. Still innocent after all these years, sketches for a social history of archaeology in Serbia. In: P. Biehl, A. Gramsch and A. Marciniak, eds. *Archaeologies of Europe. History, methods and theories / Archäologien Europas. Geschichte, methoden und theorien*. Tübinger Archäologische Taschenbücher 4. Münster: Waxman, pp. 309–22.

Babić, S. 2014. Identity, integration, power relations and the study of the European Iron Age. In: C.N. Popa and S. Stoddart, eds. *Fingerprinting the Iron Age, integrating South-Eastern Europe into the debate*. Oxford: Oxbow Books, pp. 284–90.

Babić, S. 2015. Theory in archaeology. In: J.D. Wright, ed. *International Encyclopedia of the Social & Behavioral Sciences*, 2nd edn, vol. 1. Oxford: Elsevier, pp. 899–904.

Babić, S. and M. Tomović, eds. 1996. *Milutin Garašanin – razgovori o arheologiji*. Belgrade: 3T.

Baird, S.F. 1876. Appendix to the report of the Secretary. In: *Annual Report of the Board of Regents of the Smithsonian Institution for 1875*. Washington, DC: US Government Printing Office, pp. 58–71.

Bandović, A. 2014. Muzejski kurs i arheologija tokom II svetskog rata u Beogradu. *Etnoantropološki problem* 9(3), pp. 629–48.

Bang [Alving, Barbro]. 1931. Doktorn – Landshövdingskan: ett besök hos en verksam dam med många järn i elden. *Idun* 1931(1). Unnumbered.

Bar-Yosef, O. and A. Mazar. 1982. Israeli archaeology. *World Archaeology* 13(3), pp. 310–25.

Barad, K. 2003. Posthumanist performativity: towards an understanding of how matter comes to matter. *Signs* 28(3), pp. 801–31.

Barbanera, M. 1998. *L'archeologia Degli Italiani*. Rome: Editori Riuniti.

Bardsley, C.W. 1996[1901]. *A Dictionary of English and Welsh Surnames: with special American instances*. Baltimore, Mary.: Genealogical Publishing Co.

Barnabei, M. and F. Delpino, eds. 1991. *Le memorie di un archeologo di Felice Barnabei*. Roma: De Luca Edizioni dArte.

Bartlett, J.R. 1848. The Progress of Ethnology, an account of recent archaeological, philological and geographical researches in various parts of the globe. *Transactions of the American Ethnological Society* 2 (appendix), pp. 1–149.

Baudou, E. 2004. *Den nordiska arkeologin: historia och tolkningar*. Stockholm: Kungl. Vitterhets historie och antikvitets akademien, Almqvist and Wiksell International.

Baudou, E. 2012a. *Oscar Montelius: om tidens återkomst och kulturens vandringar*. Vitterhets historie and antikvitets akademien in collaboration with Atlantis. Stockholm: Kungl.

Baudou, E. 2012b. Forskarbiografi och vetenskapshistoria: Montelius och Hildebrands liv och verk. *Att återupptäcka det glömda: aktuell forskning om forntidens förflutna i Norden*. Lund: Institutionen för arkeologi och antikens historia, Lunds Universitet. pp. 179–97.

Bažant, J. 1993. The case of the talkative connoisseur. *Eirene* 29, pp. 84–107.

Benac, A. 1964. Prediliri, protoiliri i prailiri. In: A. Benac, ed. *Simpozijum o teritorijalnom i hronološkom razgraničenju Ilira u praistorijsko doba*. Sarajevo: ANUBIH, pp. 59–94.

Benac, A. 1987. Etnogenetska pitanja. O etničkim zajednicama željeznog doba u Jugoslaviji. In: A. Benac, ed. *Praistorija Jugoslavenskih Zemalja* V: *željezno doba*. Sarajevo: Akademija nauka i umjetnosti Bosne i Hercegovine, Centar za balkanološka ispitivanja, pp. 737–802.

Bencivenni, M., R. Dalla Negra and P. Grifoni. 1987. *Monumenti e istituzioni*. Vol. I. *La nascita del servizio di tutela dei monumenti in Italia 1860–1880*. Florence: Alinea.

Bencivenni, M., R. Dalla Negra and P. Grifoni. 1992. *Monumenti e istituzioni*. Vol. II. *Il decollo e la riforma del servizio di tutela dei monumenti in Italia. 1880–1915*. Florence: Alinea.

Beneš, E. 1925. The problem of the small nations after the World War. *Slavonic Review* 4(11), pp. 257–77.

Berg, I. 2016. *Kalaureia 1894: a cultural history of the first Swedish excavation in Greece*. Stockholm: Stockholms Universitet.

Berghahn, V. and S. Lässig. 2008. *Biography between Structure and Agency: Central European lives in international historiography*. New York: Berghahn Books.

Besnard, P. 1983. The *Année sociologique* team. In: P. Besnard, ed. *The Sociological Domain: The Durkheimians and the founding of French sociology*. Cambridge: Cambridge University Press, pp. 11–39.

Bieder, R.E. 1986. *Science Encounters the Indian, 1820–1880*. Norman: University of Oklahoma Press.

Bierbrier, M.L., ed. 2012 *Who Was Who in Egyptology*, 4th edn. London: Egypt Exploration Society.

Birch, T., ed. 1772. *The Works of the Honourable Robert Boyle in six volumes: to which is prefixed the life of the author*. Vol. 6. London: privately printed.

Bissing, F.W. von. 1907. Adolf Furtwängler. *Münchner neueste Nachrichten*. 14 October, pp. 3–4.

Blanck, H. 2004. Helbig, Wolfgang. *Dizionario biografico degli Italiani*, 61, Rome: Instituto dell'Enciclopedia Italiana, pp. 670–73.

Boardman, J. 2006. *A History of Greek Vases: potters, painters and pictures*. London: Thames and Hudson.

Bokholm, S. 2000. En kvinnoröst i manssamhället: Agda Montelius (1850–1920). Stockholm: Stockholmia förlag.

Bon, F. and A. Bon. 1957. Les timbres amphoriques de Thasos. *Études thasiennes* 4. Paris – Athens.

Boon, H.N. 1989. Het Nederlands Instituut na de Tweede Wereldoorlog. *Mededelingen van het Nederlands Instituut te Rome* 49, pp. 60–76.

Borsari, L. 1886. Cerveteri. Nota del sig. Luigi Borsari. *Notizie degli Scavi di Antichità*, pp. 38–9.

Boué, A. 1840. *La Turquie d'Europe*. Paris: Arthus Bertrand.

Bourdieu, P. 1984. *Distinction: a social critique of the judgement of taste*. Cambridge, Mass.: Harvard University Press.

Bourdieu, P. 2004. *Science of Science and Reflexivity*. Cambridge: Polity Press.

Bourdieu, P. 2005. *The Social Structure of the Economy*. Cambridge: Polity Press.

Bourdin, S. and E. Nicoud. 2013. Archaeological research in foreign institutes in Rome. In: M.P. Geurmandi and K.S. Rossenbach, eds. *Twenty Years after Malta: preventive archaeology in Europe and in Italy*. Istituto per i Beni Artistici Culturali e Naturali Regione Emilia-Romagna, pp. 141–52.

Bouzek, J. 1980. The course of the Czechoslovak expedition. In: J. Bouzek et al., eds. *The Results of the Czechoslovak Expedition, Kyme II*. Prague: Univerzita Karlova, pp. 21–5.

Boyd-Dawkins, W. 1871. British bears and wolves. *The Popular Science Review* 10, pp. 241–53.

Boyle, R. 1675. *Some considerations about the reconcileableness of reason and religion*. London: T.N.

Bradley, M., ed. 2010. *Classics and Imperialism in the British Empire*. New York: Oxford University Press.

Breasted, C. 1943. *Pioneer to the Past: the story of James Henry Breasted, archaeologist, told by his son Charles Breasted*. New York: Charles Scribners Sons.

Breasted, J.H. 1897. Review of G.C. Maspero, *The Struggle of Nations*. *The Dial* 22, pp. 282–4.

Bridge, F.R. 2002. *From Sadowa to Sarajevo: the foreign policy of Austria-Hungary, 1866–1914*. London and Boston: Routledge and Kegan Paul.

Britton, J. 1845. *Memoir of John Aubrey, FRS, embracing his auto-biographical sketches, a brief review of his personal and literary merits, and an account of his works; with extracts from his correspondence, anecdotes of some of his contemporaries, and of the times in which he lived*. London: J.B. Nichols and Son.

Brooks, C.S. 1928. *Roads to the North*. New York: Harcourt Brace and Company.

Brorson, S. and H. Andersen. 2001. Stabilizing and changing phenomenal worlds: Ludwik Fleck and Thomas Kuhn on scientific literature. *Journal for General Philosophy of Science* 32, pp. 109–29.

Browne, J. 2003. *Charles Darwin: a biography*. Vol. 2: *The power of place*. Princeton, NJ: Princeton University Press.

Browne, J. 2014. Corresponding naturalists. In: B. Lightman and M.S. Reidy, eds. *The Age of Scientific Naturalism: Tyndall and his contemporaries*. London: Pickering and Chatto, pp. 157–69.

Brughmans, T. 2013. Thinking through networks: a review of formal network methods in archaeology. *Journal of Archaeological Method and Theory* 20, pp. 623–62.

Brughmans, T. 2014. The roots and shoots of archaeological network analysis: a citation analysis and review of the archaeological use of formal network methods. *Archaeological Review from Cambridge* 29(1), pp. 18–41.

Bulle, H. 1907. Adolf Furtwängler 1853–1907. *Beilage zu Allgemeine zeitschrift*, 23–24 October, pp. 33–5.

Bulletin of The Museum of Far Eastern Antiquities, 1932. Bulletin No 4. Stockholm: Östasiatiska museet. pp. v–viii.

Burdett, O. and E.H. Goddard. 1941. *Edward Perry Warren: the biography of a connoisseur*. London: Christophers.

Burl, A. 2000. *The Stone Circles of Britain, Ireland and Brittany*. New Haven, Conn.: Yale University Press.

Burl, A. 2002. *Prehistoric Avebury*. New Haven, Conn.: Yale University Press.

Burl, A. 2010. *John Aubrey and Stone Circles: Britain's first archaeologist from Avebury to Stonehenge*. Stroud: Amberley.

Burns, J.C. 2008. Networking Ohio Valley archaeology in the 1880s: The social dynamics of Peabody and Smithsonian centralization. *Histories of Anthropology* 4, pp. 1–33.

Calder, W.M. III. 1996. Adolf Furtwängler. In: N.T. de Grummond, ed. *Encyclopedia of the History of Classical Archaeology*. London and Westport, Conn.: Greenwood Press, pp. 475–6.

Callon, M. 1986. The sociology of an actor-network: the case of the electric vehicle. In: M. Callon, J. Law and A. Rip, eds. *Mapping the Dynamics of Science and Technology: sociology of science in the real world*. London: MacMillan Press, pp. 19–34.

Callon, M. and B. Latour. 1981. Unscrewing the big leviathan: how actors macro-structure reality and how sociologists help them do so. In: K. Knorr and A. Cicourel, eds. *Advances in Social Theory and Methodology*. London: Routledge and Kegan Paul, pp. 277–303.

Callon, M., J. Law and A. Rip, eds. 1986. *Mapping the Dynamics of Science and Technology: sociology of science in the real world*. London: MacMillan Press.

Camden, W. 1722. *Britannia: or a chorographical description of Great Britain and Ireland together with the adjacent islands*. London: Mary Matthews.

Carruthers, W., ed. 2014. *Histories of Egyptology: interdisciplinary measures*. London: Routledge.

Ceserani, G. 2012. *Italy's Lost Greece: Magna Graecia and the making of modern archaeology*. New York: Oxford University Press.

Challis, D. 2013. *The Archaeology of Race: the eugenic ideas of Francis Galton and Flinders Petrie*. London: Bloomsbury.

Chase, G.H. 1950. *Greek and Roman Antiquities. A guide to the classical collection*. Boston, Mass.: Museum of Fine Arts.

Chong, A., R. Lingner and C. Zahn. 2003. *Eye of the Beholder. Masterpieces from Isabella Stewart Gardner Museum*. Boston: Isabella Stewart Gardner Museum and Beacon Press.

Christenson, A.L. 1989. *Tracing Archeology's Past: the historiography of archaeology*. Carbondale: Southern Illinois University.

Church, J.E. 1908. Adolf Furtwängler: Artist, archaeologist, professor. *University of Nevada Studies* 1–2, pp. 61–6.

CIA. 1905. *Comptes rendus du Congrès International d'Archéologie*, Ier session. Athens: Imprimerie Hestia.

Cleal, R. M. J. and R. Montague. 2001. Neolithic and early Bronze Age. In: A. Chadburn and M. Pomeroy-Kellinger, eds. *Archaeological research agenda for the Avebury world heritage site*. Salisbury: Avebury Archaeological and Historical Research Group, pp. 8–19.

Cohen, G.M. and M.S. Joukowsky, eds. 2004. *Breaking Ground: pioneering women archaeologists*. Ann Arbor: University of Michigan Press.

Cohen, R.S. and T. Schnelle. 1986. Introduction. In: R.S. Cohen and T. Schnelle, eds. *Cognition and Fact, materials on Ludwik Fleck*. Boston Studies in the Philosophy of Science 87. Dordrecht: Springer, pp. ix–xxxiii.

Colt-Hoare, R. 1821. *The Ancient History of North Wiltshire*. London: W. Bulmer and Co.

Condé, M. and M. Salomon. 2016. Dossier: Ludwik Fleck's theory of thought styles and thought-collectives – translations and receptions. *Transversal, International Journal for the Historiography of Science* 1, pp. 3–115.

Conkey, M.W. and J.M. Gero. 1997. Programme to practice: gender and feminism in archaeology. *Annual Review of Anthropology* 26, pp. 411–37.

Cools, H. and H. de Valk. 2004. *Institutum Neerlandicum MCMIV–MMIV. Honderd jaar Nederlands Instituut te Rome*. Hilversum: Uitgeverij Verloren.

Coppola, M.R. 2009. Il caso di un furto di monete nel cantiere del monumento a Vittorio Emanuele II: mercato antiquario, legislazione di tutela, grandi opere per Roma Capitale. *Studi Romani* 57, pp. 198–218.

Cunnington, M.E. 1933. Wiltshire in pagan Saxon times. *Wiltshire Archaeological and Natural History Magazine* 48, pp. 147–75.

Cunnington, M.E. n.d. *Avebury: a guide to the circles, the church, the manor house etc, Silbury Hill*. Devizes: C.H. Woodward.

Curta, F. 2013. With brotherly love, the Czech beginnings of medieval archaeology in Bulgaria and Europe. In: P.J. Geary and G. Klaniczay, eds. *Manufacturing Middle Ages: entangled history of Medievalism in nineteenth-century Europe*. Leiden: Brill, pp. 377–96.

Curtius, L. 1950. *Deutsche und antike welt: lebenserinnerungen*. Stuttgart: Deustche Verlags-Anstalt.

Curtius, L. 1958. *Torso: verstreuten und nachgelassene schriften*. Stuttgart: Deutsche Verlags-Anstalt, pp. 213–24 [reprint of Curtius, L. 1935. Adolf Furtwängler. *Badische Biographien* 6, pp. 672–85.]

Cvjetićanin, T. 2011. Felix Kanitz und das antike Erbe in Serbien. In: Đ. Kostić, ed. *Balkanbilder von Felix Kanitz / Слике са Балкана Феликса Каница*. Belgrade: Nationalmuseum, pp. 145–59.

Dally, O. 2017. Zur archäologie der fotographie: ein Beitrag zu Abbildungspraktiken der zweiten Hälfte des 19. und des frühen 20. Jahrhunderts. Winckelmann-Programm der Archäologischen Gesellschaft zu Berlin, 143. Berlin: De Gruyter.

Damilano, R. 2014. Palma di Cesnola, Luigi. *Dizionario Biografico degli Italiani*, 80, Rome: Instituto dell'Enciclopedia Italiana, pp. 586–8.

Daniel, G. 1950. *A Hundred Years of Archaeology*. London: Duckworth.

Daniel, G., ed. 1978. *The Illustrated Encyclopedia of Archaeology*. London: Macmillan.

Daniel, G. 1981a. *A Short History of Archaeology*. London: Thames and Hudson.

Daniel, G. 1981b. *Towards a History of Archaeology: being the papers read at the first Conference on the History of Archaeology in Aarhus, 29 August–2 September 1978*. London: Thames and Hudson.

David, M. 2002. Urban planning and archaeology in the Roman capital. In: M. David, ed. *Ruins of Ancient Rome. The drawings of French architects who won the Prix de Rome*. Los Angeles: The J. Paul Getty Museum, pp. 36–51.

Davies, N. 1996. *Europe – a History*. New York: Oxford University Press.

Davis, J.B. and J. Thurnam. 1865. *Crania Britannia: Delineations and descriptions of the skulls of the original and early inhabitants of the British Islands with notices of their other remains*. London: printed for the subscribers.

Day, J. 2006. *The Mummy's Curse: Mummymania in the English-speaking world*. London: Routledge.

De Puma, R.D. 1994. Ancient art at the Art Institute of Chicago. *The Art Institute of Chicago Museum Studies* 20(1), pp. 54–61.

De Puma, R.D. 1996. Frothingham, Arthur Lincoln Jr. In: N. Thomson de Grummond, ed. *Encyclopedia of the History of Classical Archaeology*, vol. 1. Westport, Conn.: Greenwood Press, p. 471.

De Simone, C. 2011. Ancora sulla fibula Praenestina (e fine). In: S. Örma and K. Sandberg, eds. *Wolfgang Helbig e la scienza dellantichità del suo tempo*. Roma: Institutum Romanum Finlandiae, pp. 217–24.

Del Collo, A.M. 2011. Cultivating taste: Henry G. Marquand's public and private contributions to advancing art in Gilded Age New York. Unpublished Masters dissertation, New York.

Derrida, J and E. Prenowitz. 1995. Archive fever: a Freudian impression. *Diacritics* 25(2), pp. 9–63.

di Gennaro, F., L. La Rocca, E. Mangani, D. Ferro, E. Formigli, M. Buonocore, G.L. Carancini, G. Colonna, C. de Simone, A. Franchi De Bellis, D.F. Maras, P. Poccetti and M. Sannibale. 2015. La *fibula Prenestina. Bullettino di Paletnologia Italiana* 99.

Díaz-Andreu, M. 2007. *A World History of Nineteenth Century Archaeology. Nationalism, colonialism and the past*. Oxford: Oxford University Press.

Díaz-Andreu, M. 2008. Revisiting the 'invisible college': José Ramón Mélida in early 20th century Spain. In: N. Schlanger and J. Nordbladh, eds. *Histories of Archaeology: archives, ancestors, practices*. Oxford: Berghahn Books, pp. 121–9.

Díaz-Andreu, M. and T.C. Champion. 1996. *Nationalism and Archaeology in Europe*. London: UCL Press.
Díaz-Andreu, M. and M. Sørensen, eds. 1998. *Excavating Women: a history of women in European archaeology*. London: Routledge.
Dostálová, R. 2004. 10 let novořeckých studií na FF Masarykovy university. *Universitas*, pp. 31–4.
Dries, M. van, C. Slappendel and S.J. van der Linde. 2010. Dutch archaeology abroad: from treasure hunting to local community engagement. In: S.J. van der Linde, ed. *European Archaeology Abroad. Global setting, comparative perspectives*. Leiden: Sidestone Press, pp. 125–56.
Drower, M. 1985. *Flinders Petrie: a life in archaeology*. Madison: University of Wisconsin Press.
Duncan, W. 1970. *The Life and Times of Vuk Stefanović Karadžić, 1787–1864: literacy, literature and national independence in Serbia*. Oxford: Clarendon Press.
Dyer, J. 1990. *Ancient Britain*. London: B.T. Batsford Ltd.
Dyson, S.L. 1998. *Ancient Marbles to America's Shores*. Philadelphia: University of Pennsylvania Press.
Dyson, S.L. 2004. *Eugénie Sellers Strong: portrait of an archaeologist*. London: Duckworth.
Dyson, S.L. 2006. *In Pursuit of Ancient Pasts: a history of Classical Archaeology in the nineteenth and twentieth centuries*. New Haven, Conn.: Yale University.
Džino, D. 2014. Constructing Illyrians: prehistoric inhabitants of the Balkan Peninsula in early modern and modern perceptions. *Balkanistica* 27, pp. 1–39.
Eichmann, K. 2008. The anthropology of science: Ludwik Fleck, et al.. In: K. Eichmann, ed. *The Network Collective: rise and fall of a scientific paradigm*. Basel: Birkhäuser, pp. 26–42.
Ellis, J.H., ed. 1887. *The Registers of Stourton, County Wilts, from 1570–1800*. London: no publisher.
Essen, C.C. van. 1950. De tegenwoordige mogelijkheden voor archeologische studie in Rome en omgeving. *Mededelingen van het Nederlands Instituut te Rome*, 26, pp. xlii–l.
Evans, C. 1989. Archaeology and modern times: Bersu's Woodbury 1938 & 1939. *Antiquity* 63, pp. 436–50.
Evans, C. 1990. 'Power on silt': towards an archaeology of the East India Company. *Antiquity* 64, pp. 643–61.
Evans, C. 1998. Historicism, chronology and straw men: situating Hawkes' 'ladder of inference'. *Antiquity* 72, pp. 398–404.
Evans, S.B. 1877. *Final report of the Ohio State Board of Centennial Managers to the General Assembly of the State of Ohio*. Columbus: Nevins and Myers.
Evans, S.B. 1880. Notes on the principal mounds in the Des Moines River Valley. *Annual report of the Smithsonian Institution for 1879*. Washington, DC: US Government Printing Office, pp. 344–9.
Fehér, G. 1932. *Kanitz Fülöp Félix, a Balkàn Kolumbusa élete és munkássága 1829–1904*. Budapest: Franklin Társulat.
Fergusson, J. 1872. *Rude Stone Monuments in all Countries: their age and uses*. London: John Murray.
Ferrua, A. 1940a. Il mitreo sotta la chiesa di Santa Prisca. *Bullettino della Commissione Archeologica Comunale di Roma* 68, pp. 59–96.

Ferrua, A. 1940b. Recenti ritrovamenti a S. Prisca. In: *Rivista di Archeologia Cristiana* 17, pp. 271–5.

Finnegan, R., ed. 2005. *Participating in the Knowledge Society: researchers beyond the university walls*. London: Palgrave MacMillan.

Fiorelli, G. 1885. *Sullordinamento del Servizio Archeologico, seconda relazione del direttore generale delle antichità e belle arti a S. E. il ministro della Istruzione pubblica*. Rome: Tipografia della Camera dei Deputati.

Fittschen, H. 1996. L'École Française d'Athènes et l'Institut archéologique Allemand. *Bulletin de Correspondance Hellénique* 120(1), pp. 487–96.

Fleck, L. 1979, 1981 [1935]. *Genesis and Development of a Scientific Fact*. English edns, trans. F. Bradley and T.J. Trenn. Chicago: University of Chicago Press.

Fleck, L. 1986 [1947]. To look, to see, to know. In: R.S. Cohen and T. Schnelle, eds. *Cognition and Fact, materials on Ludwik Fleck*. Boston Studies in the Philosophy of Science 87. Dordrecht: Springer, pp. 129–51.

Foster, J.W. 1873. *Pre-historic Races of the United States of America*. Chicago: S. C. Griggs and Co.

Foster, J. 1891. *Alumni Oxoniensis: the members of the University of Oxford 1500–1714*. Oxford: James Parker and Co.

Fowler, C. and O.J.T. Harris. 2015. Enduring relations: exploring a paradox of new materialism. *Journal of Material Culture* 20(2), pp. 127–48.

Fowles, J., ed. 1980. *John Aubrey's Monumenta Britannica*. Vol. 1. Port Sherbourne: DPC.

Fowles, J. and B. Brukoff. 1980. *The Enigma of Stonehenge*. London: Jonathan Cape.

Franchi De Bellis, A. 2011. La *fibula prenestina*, Margherita Guarducci e Wolfgang Helbig, presunto falsario. In: S. Örma and K. Sandberg, eds. *Wolfgang Helbig e la scienza dellantichità del suo tempo*. Rome: Institutum Romanum Finlandiae, pp. 169–204.

Fridh-Haneson, B-M. and I. Haglund, eds. 2004. *Förbjuden frukt på kunskapens träd: Kvinnliga akademiker under 100 år*. Stockholm: Atlantis.

Fried, M. 1977. Review. *The Religion of the Chinese People* by Marcel Granet and Maurice Freedman. *The China Quarterly* 69, pp. 157–60.

Frolíková, A. 1987. Stipendijní cesty do Řecka na přelomu století. *Listy filologické* 110(2), pp. 121–3.

Frolíková, A. and P. Oliva. 2013. Eirene. Studia Graeca et Latina. Interviewed by J. Čechvala, *Akademický bulletin*.

Furtwängler, A. 1874. *Eros in der Vasenmalerei*. Munich: Ackermann.

Furtwängler, A. 1875. Le Musée Fol. *Jenaer Literaturzeitung* 4, p. 64.

Furtwängler, A. 1883–87. *Der Sammlung Sabouroff: Kunstdenkmäler aus Griechenland*. Berlin: Asher.

Furtwängler, A. 1885. *Beschreibung der Vasensammlung im Antiquarium, Königlische Museen zu Berlin*. Berlin: Spemann.

Furtwängler, A. 1886. *Mykenische Vasen: Vorhellenistische Thongefässe aus dem Gebiete des Mittelmeere*. Berlin: Asher.

Furtwängler, A. 1890. *Olympia: die Ergebnisse der von dem deutschen Reich veranstalteten Ausgrabung*. Vol. 4, *Die Bronzen und übrigen kleineren Funde von Olympia*. Berlin: Asher.

Furtwängler, A. 1893. *Meisterwerke der Griechischen Plastik: kunstgeschichtliche Untersuchungen*. Berlin and Leipzig: Giesecke and Devrient.

Furtwängler, A. 1896. *Beschreibung der geschittenen Steine im Antiquarium, Königliche Museen zu Berlin*. Berlin: Spemann.

Furtwängler, A. 1900. *Die antiken Gemmen: Geschichte der Steinschneidekunst im klassischen Altertum*. Berlin and Leipzig: Giesecke and Devrient.

Furtwängler, A. 1965. *Briefe aus dem Bonner privatdozentenjahr 1879/80 und der Zeit seiner Tätigkeit an den Berliner Museen 1880–1894*, ed. A. Greifenhagen. Stuttgart: Kolhammer.

Furtwängler, A. and G. Loeschcke. 1879. *Mykenische Tongefässe*. Berlin: Asher.

Furtwängler, A. and G. Loeschcke. 1886. *Mykenische Vasen: Vorhellenistische Thongefässe aus dem Gebiete des Mittelmeere*. Berlin: Asher.

Furtwängler, A.E. 2005. Aspekte einer unkonventionellen Persönlichkeit in privatem Umfeld. In: V.M. Strocka, ed. *Meisterwerke: internationales Symposion anläßlich des 150. Geburtstages von Adolf Furtwängler*, Freiburg im-Breisgau 13 June–3 July 2003. Munich: Hirmer, pp. 9–19.

Fyfe, A. and B. Lightman, eds. 2007. *Science in the Marketplace: nineteenth-century sites and experiences*. Chicago: University of Chicago Press.

Gačić, D. 2005. (Miodrag Grbić 1901–1969) život i delo/*(Миодраг Грбић 1901–1969) живот и дело*. Sremski Karlovci: Zavičajna zbirka Sremski Karlovci.

Garašanin, M. 1950. Review of V. Gordon Childe, *The Dawn of European Civilisation*, fourth edition revised and reset. *Starinar/Старинар n.s.* I, pp. 256–7.

Garašanin, M. 1959a. Review of V. Gordon Childe, *The Dawn of European Civilisation*, sixth edition revised and reset. *Starinar/Старинар n.s.* IX–X, pp. 392–3.

Garašanin, M. 1959b. Obituary: V. Gordon Childe. *Starinar/Старинар n.s.* IX–X, p. 446.

Garašanin, M. 1964. Problem kontinuiteta u arheologiji. *Materijali* I, pp. 9–45.

Garašanin, M. 1988. Nastanak i poreklo Ilira/Настанак и порекло Илира. In: M. Garašanin, ed. *Iliri i Albanci/Илири и Албанци*. Belgrade: Srpska akademija nauka i umetnosti, pp. 9–144.

Gardner, P. 1907. Adolf Furtwängler. *Classical Review* 21, pp. 251–3.

Gatti, G. 1887. Trovamenti riguardanti la topografia e la epigrafia urbana. *Bullettino della Commissione Archeologica Municipale di Roma*, pp. 173–91.

Gavela, B. 1971. O arheološkom identifikovanju ilirskog etnosa. *Materijali* VII, pp. 21–39.

Gell, A. 1998. *Art and Agency: an anthropological theory*. Oxford: Clarendon Press.

Gellner, E. 1994. *Encounters with Nationalism*. Cambridge: Blackwell.

Gero, J.M. and M.W. Conkey. 1991. *Engendering Archaeology: women and prehistory*. London: Wiley-Blackwell.

Giberti, B. 2002. *Designing the Centennial: a history of the 1876 International Exhibition in Philadelphia*. Lexington: University Press of Kentucky.

Gierow, P.G. 1995. Oscar Montelius et la civilisation primitive en Italie depuis l'introduction des métaux. In: P. Åström, ed. *Oscar Montelius 150 years*, Stockholm: Kungl. Vitterhets historie och antikvitets akademien, pp. 67–75.

Gillberg, Å. and O. Jensen. 2007. Processes of professionalization and marginalization: a constructivist study of field practices in Sweden 1807–1910. In: E. Bentz and E. Rudebeck, eds with a contribution by P. Lekber. *Arkeologins många roller och paktiker* 1, pp. 9–31.

Glänzel, W. and A. Schubert. 2005. Analysing scientific networks through co-authorship. In: H.F. Moed, W. Glänzel and U. Schmoch, eds. *Handbook of Quantitative Science and Technology Research: the use of publication and patent statistics in studies of S&T systems*. New York: Kluwer Academic Publishers, pp. 257–6.

Gmür, M. 2006. Co-citation analysis and the search for invisible colleges: a methodological evaluation. *Scientometrics* 57(1), pp. 27–57.

Goffman, E. 1959. *The Presentation of Self in Everyday Life*. Garden City, NJ: Doubleday.

Goldstein, D. 1994. Yours for science: The Smithsonian Institution's correspondents and the shape of scientific community in nineteenth century America. *Isis* 85(4), pp. 573–99.

Goldstein, D. 2008. Outposts of science: The knowledge trade and the expansion of scientific community in post-Civil War America. *Isis* 99(3), pp. 519–46.

Goldsworthy, V. 2006. The Balkans in nineteenth-century British travel writing. In: T. Youngs, ed. *Travel Writing in the Nineteenth Century: filling the blank spaces*. London: Anthem Press, pp. 19–35.

Goodman, D. 1994. *The Republic of Letters: a cultural history of the French Enlightenment*. Ithaca, NY: Cornell University Press.

Gori, M. 2014. Fabricating identity from ancient shards: memory construction and cultural apropriation in the New Macedonian question. *Hungarian Historical Review* 3(2), p. 285–311.

Gough, R., ed. 1789. *Britannica*. London: John Nichols.

Gran-Aymerich, E. 1998. *Naissance de l'archéologie moderne: 1798–1945*. Paris: CNRS.

Gran-Aymerich, E. 2001. *Dictionnaire biographique d'archéologie: 1798–1945*. Paris: CNRS.

Gran-Aymerich, E. 2007. *Les chercheurs de passé: 1798–1945*. Paris: CNRS.

Granet, M. 1922. *La religion des Chinois*. Paris: Presses Universitaires de France.

Granet, M. 1977 [1975]. *The Religion of the Chinese People*, trans. and intro. M. Freedman. London: Harper and Row.

Gräslund, B. 1974. Relative datering: om kronologisk metod i nordisk arkeologi – [Relative chronology: on dating methods in Scandinavian archaeology]. Unpublished dissertation, Uppsala University.

Green, M. 1989. *The Mount Vernon Street Warrens*. New York: Charles Scribner's Sons.

Green, S. 1981. *Prehistorian: A biography of V. Gordon Childe*. Bradford-on-Avon: Moonraker Press.

Greifenhagen, A. 1965. Der archäologe Adolf Furtwängler und der Berliner Museen, 1880–1884. *Jahrbuch der Stiftung Preußischer Kulturbesitz* 3, pp. 136–9.

Grinsell, L.V. 1936. *The Ancient Burial Mounds of England*. London: Methuen.

Grinsell, L.V. 1976. *Folklore of Prehistoric Sites in Britain*. London: David and Charles.

Groote, M. von. 1912. *Ägineten und archäologen: Ein Kritik*. Strasbourg: Heitz and Mündel.

Guarducci, M. 1980. La cosiddetta *fibula Prenestina*. Antiquari, eruditi e falsari nella Roma dell Ottocento. *Memorie dell Accademia Nazionale dei Lincei, cl. di Scienze Morali, Storiche e Filologiche* 24, pp. 415–574.
Guarducci, M. 1984. La cosiddetta *fibula Prenestina*: elementi nuovi. *Memorie dell Accademia Nazionale dei Lincei, cl. di Scienze Morali, Storiche e Filologiche* 28, pp. 125–77.
Guidi, A. 2008. Italian prehistoric archaeology in the international context. *Fragmenta* 2, pp. 109–23.
Guidott, T. 1708. *An apology for the bath, being an answer to a late enquiry into the right use and abuses of the baths in England so far as may concern the hot waters of the Bath in the County of Somerset with some reflections on fresh cold bathing, Bathing in sea water and dipping in baptism in a letter to a friend.* London: no publisher.
de Haan, N., M. Eickhoff and M. Schwegman, eds. 2008. *Archaeology and National Identity in Italy and Europe, 1800–1950. Fragmenta.* Journal of the Royal Netherlands Institute in Rome. Turnhout: Brepols.
Haitsma Mulier, E. 1991. In memoriam. Adriaan Luijdjens, een Romeinse Nederlander. *Incontri* 6, pp. 159–63.
Hall, M. 2000. *Archaeology and the Modern World: colonial transcripts in South Africa and Chesapeake.* London: Routledge.
Hallind, K. 2004. Lärda kvinnor i Lund. In: B.M. Fridh-Haneson and I. Haglund, eds. *Förbjuden frukt på kunskapens träd: kvinnliga akademiker under 100 år.* Stockholm: Atlantis, pp. 183–217.
Halpern, J.M. 2014. Transitions in southeast European Studies In: C. Promitzer, S. Gruber and H. Heppner, eds. *Southeast European Studies in a Globalizing World.* Münster: LIT Verlag, pp. 9–29.
Ham-Milovanović, L. 2009 Naziv i ustrojstvo Muzeja kneza Pavla/Назив и устројство Музеја кнеза Павла. In: T. Cvjetićanin, ed. *Muzej kneza Pavla/Музеј кнеза Павла.* Belgrade: Narodni muzej u Beogradu. pp. 90–123.
Hamilakis, Y. 2007. *The Nation and its Ruins: Antiquity, archaeology, and national imagination in Greece.* New York: Oxford University Press.
Hamilakis, Y. 2010. Review of Schlanger and Nordbladh, eds. *Archives, Ancestors, Practices. Antiquity* 84, pp. 893–4.
Hankey, J. 2001. *A Passion for Egypt: Arthur Weigall, Tutankhamun and the curse of the pharaohs.* London: I.B. Tauris.
Hansson, U.R. 2008. 'Arkeologins Linné': Adolf Furtwängler som arkeologisk storstipendiat i Rom 1876–1878. *Romhorisont* 49, pp. 19–23.
Haraway, D. 1992. *Primate Visions: Gender, race and nature in the world of modern science.* London: Verso.
Harding, S. 1991. *Whose Science? Whose Knowledge? Thinking from women's lives.* Milton Keynes: Open University Press.
Härke, H. 2000. *Archaeology, Ideology, and Society: the German experience.* Frankfurt am Main: P. Lang.
Hauser, F. 1908. Adolf Furtwängler. *Süddeutsche Monatshefte* 5(4), pp. 461–72.
Haven, S.F. 1856. Archaeology of the United States, or, sketches historical and bibliographical, of the progress of information and opinion respecting vestiges of antiquity in the United States. *Smithsonian Contributions to Knowledge* 8. Washington, DC.

Havlíková, L. 1999. Česká byzantologie a slovanský ústav. *Slavia* 68(3–4), pp. 442–51.

Havránek, B. 2009. The education of Czechs and Slovaks under foreign domination. In: *University, Historiography, Politics*. Prague: Nakladatelství Karolinum, pp. 43–65.

Helbig, W. 1887. Sopra una fibula doro trovata presso Palestrina. *Römische Mitteilungen* 2, pp. 36–9.

Helbig, W. 1891. *Führer durch die őffentlichen Sammlungen klassischer Altertümer in Rom*. Leipzig: Teubner.

Helbig, W. 1927. Eine Skizze meines wissenschaftlichen Bildungsganges. In: F. Poulsen, ed. *Das Helbig Museum der Ny Carlsberg Glyptothek. Beschreibung der etruskischen Sammlung*. Copenhagen: Gedruckt bei Neilsen and Lydiche, pp. iii–xiv.

Henry, J. 1876. Report of Joseph Henry, Secretary, for the year 1875. In: *Annual Report of the Board of Regents of the Smithsonian Institution for 1875*. Washington, DC: US Government Printing Office, pp. 7–57.

Henry, J. 1996. *The Papers of Joseph Henry*, Vol. 7, *January 1847–1849: the Smithsonian years*. Washington, DC: Smithsonian Institution Press.

Henson, P.M. 2000. Spencer Baird's dream: a US national museum. In: M.T. Ghiselin and A.E. Leviton, eds. *Cultures and Institutions of Natural History: essays in the history and philosophy of science*. San Francisco: California Academy of Sciences, pp. 101–26.

Heres, T.L. 1989. Geschiedenis van de archeologie aan het Nederlands Instituut, 1904–1965. *Mededelingen van het Nederlands Instituut te Rome* 49, pp. 77–85.

Hicks, D. and M.C. Beaudry. 2006. *The Cambridge Companion to Historical Archaeology*. Cambridge: Cambridge University Press.

Higgins, G. 2007 [1829]. *The Celtic Druids*. New York: Cosimo Classics.

Higonnet, A. 2009. *A Museum of One's Own. Private collecting, public gift*. Pittsburgh, Penn.: Periscope Publishing; New York: Prestel Publishing.

Hildebrand, H. 1872. *Den arkeologiska kongressen i Bologna: Berättelse*. Stockholm: Kungl. Vitterhets historie och antikvitets akademien.

Hilliard, C. 2010. *The Prime Mover: Charles L. Hutchinson and the making of the Art Institute of Chicago*. Chicago: Art Institute of Chicago.

Hinsley, C.M., Jr. 1981. *The Smithsonian and the American Indian: making a moral anthropology in Victorian America*. Washington, DC: Smithsonian Institution Press.

Hnilica, J. 2009. *Francouzský Institut v Praze 1920–1951, Mezi Vzděláním a Propagandou*. Prague: Nakladatelství Karolinum.

Hodder, I. 1986. *Reading the Past*. Cambridge: Cambridge University Press.

Hofter, M.R. 2003. Zeit des erwachsens: Adolf Furtwänglers Briefe aus Freiburg an Heinrich Brunn. In: M. Flashar and J. Wohlfeil, eds. *Adolf Furtwängler: der Archäologe*. Schriften der Archäologischen Sammlung Freiburg. Munich: Biering and Brinkmann, pp. 21–46.

Holy, L. 1996. *The Little Czech and the Great Czech Nation*. Cambridge: Cambridge University Press.

Horel, C. 2011. Die familie Kanitz in óbuda 1829–1848. Ein Kontextualisierungsversuch. In Đ. Kostić, ed. *Balkanbilder von Felix Kanitz / Слике са Балкана Феликса Каница*. Belgrade: Nationalmuseum, pp. 7–21.

Hroch, M. 1985. *Social Preconditions of National Revival in Europe: a comparative analysis of the social composition of patriotic groups among the smaller European nations*, trans. B. Fowkes. New York: Cambridge University Press.

Hubert, H. 1905. Étude sommaire de la représentation du temps dans la religion et la magie. *Annuaire de l'École Pratique des Hautes Études, Section des Sciences Religieuses* pp. 1–39.

Hubert, H. 1919. Preface. In: S.Z. Czarnowski, ed. *Le culte des héros et ses conditions sociales*. Paris: Alcan, pp. i–xciv.

Hubert, H. 1987 [1934]. *The Greatness and Decline of the Celts*. M. Mauss, R. Lantier and J. Marx, eds. London: Constable and Company.

Isambert, F.A. 1983. At the frontier of folklore and sociology: Hubert, Hertz and Czarnowski, founders of a sociology of folk religion. In: P. Besnard, ed. *The Sociological Domain: the Durkheimians and the founding of French sociology*. Cambridge: Cambridge University Press, pp. 152–76.

James, T.G.H. 1992. *Howard Carter: the path to Tutankhamun*. London: Kegan Paul.

Jandolo, A. 1935. *Memorie di un antiquario*. Milan: Ceschina.

Janssen, R.M. 1992. *The First Hundred Years: Egyptology at University College London, 1892–1992*. London: University College London Press.

Jeffreys, D., ed. 2003. *Views of Ancient Egypt since Napoleon Bonaparte: imperialism, colonialism, and modern appropriations*. London: UCL Press.

Johansen, F. 1994. *Catalogue Roman portraits, Ny Carlsberg Glyptotek*, vol. 1. Copenhagen: Ny Carlsberg Glyptotek.

Johnston, W.R. 1999. *William and Henry Walters, the reticent collectors*. Baltimore, Mary.: Johns Hopkins University Press and Walters Art Gallery.

Kains-Jackson, C.P. 1880. *Our Ancient Monuments and the Lands around them*. London: Elliot Stock.

Kanitz, F. 1861. *Die römischen Funde in Serbien*. Vienna: Kais. Akademie der Wissenschaften.

Kanitz, F. 1862. *Serbiens Byzantinische Monumente, gezeichnet und beschrieben*. Vienna: K.K. Hof- und Staatsdrukerei.

Kanitz, F. 1868. *Serbien. historich-etnographische Reisestudien aus den Jahren 1859–1868*. Leipzig: Verlagsbuchhandlung von Hermann Fries.

Kanitz, F. 1892. *Romische Studien in Serbien; der Donau-Grenzwall, das Strassennetz, die Stadte, Castelle, Denkmale, Thermen und Bergwerke zur Romerzeit im Konigreiche Serbien*. Vienna: Akademie der Wissenschaften.

Kanitz, F. 1904. *Das Königreich Serbien und das Serbenvolk von der Römerzeit bis zur Gegenwart*. Vol. 1: *Land und Bevölkerung*. Leipzig: Verlag von Bernh. Meyer.

Kekulé, R. 1895. Review of Furtwängler 1893. [*Meisterwerke der griechischen Plastik*]. *Göttingische gelehrte anzeigen* 157, pp. 625–43.

Kekulé, R. 1908. Adolf Furtwängler. *Archäologischer Anzeiger*, pp. 39–41.

Kelly, J.E. 2002. Charles Rau: developments in the career of a nineteenth-century German-American archaeologist. In: D.L. Browman and S. Williams, eds. *New Perspectives on the Origins of Americanist Archaeology*. Tuscaloosa: The University of Alabama Press, pp. 117–32.

Ketelaar, E. 2001. Tacit narratives: the meaning of archives. *Archival Science* 1, pp. 131–41.

Kirby, T.E., ed. 1903. *Catalogue of the Art and Literary Property collected by the late Henry G. Marquand*. New York: The American Art Association.
Klejn, L.S. 2012. *Soviet Archaeology: schools, trends, history*. English edn, trans. R. Ireland and K. Windle. Oxford: Oxford University Press.
Klindt-Jensen, O. 1975. *A History of Scandinavian Archaeology*. London: Thames and Hudson.
Kočí, J. 1978. *České národní obrození*. Prague: Nakladatelství Svoboda.
Kohl, P.L. and C.P. Fawcett. 1995. *Nationalism, Politics, and the Practice of Archaeology*. Cambridge: Cambridge University Press.
Kohlstedt, S.G. 1971. A step towards scientific identity in the United States: the failure of the National Institute, 1844. *Isis* 62(3), pp. 339–62.
Kostić, Đ.S. 2011. *Dunavski limes Feliksa Kanica*. Belgrade: Arheološki institut.
Kruijer, L., J. Hilbrants, J. Pelgrom and M. Taviani. 2018. The excavations underneath the garden of the church of Santa Prisca (1964–1966). Unlocking the legacy data of the first Dutch excavation in Italy. *BABESCH* 93, pp. 235–54.
Kuhn, T.S. 1970. *The Structure of Scientific Revolutions*. Chicago: University of Chicago Press.
Kuzmanović, Z. and V.D. Mihajlović. 2015. Roman emperors and identity constructions in modern Serbia. *Identities: global studies in culture and power* 22(4), pp. 416–32.
Lanciani, R.A. 1888. *Ancient Rome in the Light of Recent Discoveries*. Boston, New York: Houghton, Mifflin and Co.
Lanciani, R.A. 1893–1901. *Forma urbis Romae*. Milan: Hoepli.
Lanciani, R.A. 1899. *The Destruction of Ancient Rome: a sketch of the history of the monuments*. New York: Macmillan and Co.
Lanciani, R.A. 1902–12. *Storia degli scavi di Roma e notizie intorno le collezioni Romane di antichità*. Rome: E. Loescher and Co.
Lapham, I.A. 1855. The antiquities of Wisconsin. *Smithsonian Contributions to Knowledge*, Vol. 7. Washington, DC.
Latour, B. 1987. *Science in Action. How to follow scientists and engineers through society*. Cambridge, Mass.: Harvard University Press.
Latour, B. 1996. On actor-network theory: a few clarifications. *Soziale Welt* 74(4), pp. 369–81.
Latour, B. 2005. *Re-assembling the Social: an introduction to actor-network theory*. Oxford: Oxford University Press.
Lavin Aronberg, M.A. 1983. *The Eye of the Tiger. The founding and development of the Department of Art and Archaeology, 1883–1923*. Princeton, NJ: Dept. of Art and Archaeology and the Art Museum, Princeton University, 1983.
Law, J. 1992. Notes on the theory of the actor-network: ordering, strategy, and heterogeneity. *Systems Practice* 5(4), pp. 379–93.
Leslie, F. 1876. *Frank Leslie's Illustrated Historical Register of the Centennial Exposition*. New York: Frank Leslie.
Levine, P. 1986. *The Amateur and the Professional: antiquarians, historians and archaeologists in Victorian England, 1838–1886*. Cambridge: Cambridge University Press.
Levkoff, M.L. 2008. *Hearst the Collector*. New York: Abrams.
Lightman, B. 2007. *Victorian Popularizers of Science: designing nature for new audiences*. Chicago: University of Chicago Press.

Liliequist, B. 2003. *Ludwik Flecks jämförande kunskapsteori.* Umeå University: Department of Philosophy and Linguistics.

Lipstadt, H. 2007. 'Life as a ride on the metro.' Pierre Bourdieu on biography and space. In: D. Arnold and J. Sofaer, eds, *Biographies and Space. Placing the Subject in Art and Architecture*, London: Routledge, pp. 35–54.

Linde, S.J. van der, ed. 2012a. *European Archaeology Abroad. Global setting, comparative perspectives.* Leiden: Sidestone Press.

Linde, S.J. van der. 2012b. *Digging Holes Abroad. An ethnography of Dutch archaeological research projects abroad.* Leiden: Leiden University Press.

Livingstone, D. 2002. Reading the heavens, planting the earth: cultures of British science. *History Workshop Journal* 54, pp. 236–41.

Livingstone, D. 2003. *Putting Science in its Place. Geographies of scientific knowledge.* Chicago and London: University of Chicago Press.

Livingstone, D. 2005. Text, talk and testimony: geographical reflections on scientific habits. An afterword. *The British Journal for the History of Science* 38, pp. 93–100.

Livingstone, D. 2007. Science, site and speech: scientific knowledge and the spaces of rhetoric. *History of the Human Sciences* 20(2), pp. 71–98.

Livingstone, D. 2010. Landscapes of knowledge. In: P. Meuseburger, D.N. Livingstone and H. Jöns, eds. *Geographies of Science.* Dordrecht: Springer, pp. 3–21.

Livingstone, D. and Withers, C., eds. 2011. *Geographies of Nineteenth-century Science.* Chicago: University of Chicago Press.

Long, W. 1858a. Abury. *Wiltshire Archaeological and Natural History Magazine* IV, pp. 307–63.

Long, W. 1858b. *Abury illustrated.* Devizes: H. Bull.

Lory, B. 2011. Felix Kanitz und Guillaume Lejean: kollegen oder rivale? In: Đ. Kostić, ed. *Balkanbilder von Felix Kanitz / Слике са Балкана Феликса Каница.* Belgrade: Nationalmuseum, pp. 69–79.

Löwy, I. 2008. Ways of seeing: Ludwik Fleck and Polish debates on the perception of reality, 1890–1947. *Studies of History and Philosophy of Science* 39(3), pp. 375–83.

Lucas, G. 2001. *Critical Approaches to Fieldwork: contemporary and historical archeological practice.* London: Routledge.

Lucas, G. 2012. *Understanding the Archaeological Record.* Cambridge: Cambridge University Press.

[Luijdjens, A.] 1952. Belangrijke opgravingen te Rome. *Algemeen Handelsblad,* 7 October, p. 1.

[Luijdjens, A.] 1958. Nederlandse archeologen volbrachten in Rome een belangrijk werk. *Algemeen Handelsblad,* 16 June, p. 3.

Luke, C. and M. Kersel. 2012. *US Archaeology and Cultural Diplomacy: soft power, hard heritage.* New York: Routledge.

Luković, M. 2011. Development of the modern Serbian state and abolishment of Ottoman agrarian relations in the 19th century. *Český lid* 98(3), pp. 281–305.

Lullies, R. 1969. Zu einer Photographie: der Archäologe Adolf Furtwängler im Kreise von Schülern und Wiener Kollegen, Wien 1905. In: P. Zazoff, ed. *Opus nobile: festschrift zum 60. geburtstag von Ulf Jantzen.* Wiesbaden: Franz Steiner Verlag, pp. 99–104.

Lundström, C. 2005. *Fruars Makt och omakt: kön, klass och kulturarv*. Umeå: Institutionen för Historiska Studier, Umeå Universitet.

MacDonald, S. and M. Rice, eds. 2003. *Consuming Ancient Egypt*. London: UCL Press.

Mach, E. von. 1907. Adolf Furtwängler. *American Anthropologist* 9, pp. 770–1.

Mackenzie, G.M. and A.P. Irby. 1877. *Travels in the Slavonic Provinces of Turkey-in-Europe*. London: Daldy, Isbister and Co.

Macura, V. 1983. *Znamení zrodu: české obrození jako kulturní typ*. Prague: Československý spisovatel.

Maddison, R.E.W. 1958. A tentative index of the correspondence of the Honourable Robert Boyle, FRS. *Notes and Records of the Royal Society of London* 13(2), pp. 128–201.

Malone, C. 1990. *Prehistoric Monuments of Avebury*. London: English Heritage.

Manco, J. 2004. The history of Acton Court. In: K. Rodwell and R. Bell, eds. *Acton Court: the evolution of an early Tudor Courtier's House*. London: English Heritage, pp. 13–42.

Mannheim, K. 1952. The problem of generations. In: P. Kecskemeti, ed. *Essays on the Sociology of Knowledge by Karl Mannheim*. New York: Routledge and Kegan Paul, pp. 276–320.

Manoff, M. 2004. Theories of the archive from across the disciplines. *portal: Libraries and the Academy* 4(1), pp. 9–25.

Marchand, S.L. 1996. *Down from Olympus: archaeology and philhellenism in Germany, 1750–1970*. Princeton, NJ: Princeton University Press.

Marchand, S.L. 2002. Adolf Furtwängler in Olympia: on excavation, the antiquarian tradition, and philhellenism in nineteenth-century Germany. In: H. Kyrieleis, ed. *Olympia 1875–2000: 125 Jahre deutsche ausgrabungen. Internazionales symposium*, Berlin 9–11 November 2000. Mainz: Philipp von Zabern, pp. 147–62.

Marchand, S.L. 2007. From antiquarian to archaeologist? Adolf Furtwängler and the problem of modern Classical Archaeology. In: P.N. Miller, ed. *Momigliano and Antiquarianism: foundations of the modern cultural sciences*. Toronto: University of Toronto Press, pp. 248–85.

Marès, A. 2004. La culture comme instrument des relations internationales: le cas franco-tchécoslovaque dans l'entre-deux guerres. In: H. Voisine-Jechová, ed. *Images de la bohême dans les lettres Françaises*. Paris: Presses de l'Université de Paris-Sorbonne, pp. 149–62.

Markusson Winkvist, H. 2003. *Som isolerade öar: de lagerkransade kvinnorna och akademin under 1900-talets första hälft*. Stockholm and Stehag: Symposion.

Masaryk, T.G. 1916. *The Problem of Small Nations in the European Crisis*. London: The Council for the Study of International Relations.

Mason, O.T. 1880. Summary of correspondence of the Smithsonian Institution previous to January 1, 1880, in response to Circular No. 316. In: *Annual Report of the Board of Regents of the Smithsonian Institution for 1879*. Washington, DC: US Government Printing Office, pp. 428–48.

Maspero, G. 1896. *The Struggle of the Nations: Egypt, Syria and Assyria*. London: Society for Promoting Christian Knowledge.

Mattaliano, E. 1975. Il movimento legislativo per la tutela delle cose di interesse artistico e storico dal 1861 al 1939. *Quaderni di Studi e Legislazione. Ricerca sui Beni Culturali* 17(1), pp. 3–89.

Mauss, M. 1983. An intellectual self-portrait. In: P. Besnard, ed. *The Sociological Domain: the Durkheimians and the founding of French sociology*. Cambridge: Cambridge University Press, pp. 139–51.

McFadden, E. 1971. *The Glitter & the Gold: a spirited account of the Metropolitan Museum of Art's first Director, the audacious and high-handed Luigi Palma di Cesnola*. New York: Dial Press.

Meskell, L. 1998. *Archaeology under Fire: nationalism, politics and heritage in the Eastern Mediterranean and Middle East*. London: Routledge.

Michel, B. 2004. La tradition des relations franco-tchèque. In: H. Voisine-Jechová, ed. *Images de la Bohême dans les lettres Françaises*. Paris: Presses de l'Université de Paris-Sorbonne, pp. 13–18.

Mihajlović, V.D. 2014. Tracing ethnicity backwards: the case of the central Balkan tribes. In: C.N. Popa and S. Stoddart, eds. *Fingerprinting the Iron Age: integrating South-Eastern Europe into the debate*. Oxford: Oxbow Books, pp. 97–107.

Mihajlović, V.V. 2016. L.F. Marsigli and the reception of Classical heritage in Serbian archaeology. Unpublished PhD dissertation, University of Belgrade, 2016 [in Serbian with English summary].

Milinković, M. 1985. Istraživanja Janka Šafarika u Rudničkom i Čačanskom okrugu i počeci arheologije u Srbiji/Истраживања Јанка Шафарика у Рудничком и Чачанском округу и почеци археологије у Србији. *Glasnik SAD* 2, pp. 74–80.

Milinković, M. 1998. Odeljenje za arheologiju/Одељење за археологију. In: Filozofski fakultet 1838–1998. Period 1963–1998/*Филозофски факултет 1838–1998. Период 1963–1998*. Belgrade: University of Belgrade, pp. 425–40.

Milosavljević, M. 2015. Koncept drugosti varvarstva i varvarizacije u srpskoj arheologiji. Unpublished PhD dissertation, Department of Archaeology, University of Belgrade.

Mitchell, T. 1991. *Colonising Egypt*. Berkeley: University of California Press.

Mizoguchi, K. 2011. *The Archaeology of Japan, from the earliest rice farming to the rise of the state*. Cambridge: Cambridge University Press.

Molinari, M.C., ed. 1994. *Il tesoro di Via Alessandrina*. Cinisello Balsamo, MI: Silvana.

Moltesen, M. 2012. *Perfect Partners. The collaboration between Carl Jacobsen and his agent in Rome Wolfgang Helbig in the formation of the Ny Carlsberg Glyptotek 1887–1914*. Copenhagen: Ny Carlsberg Glyptotek.

Moltesen, M. forthcoming. Paul Arndt, Classical archaeologist and agent in the ancient art trade, and his collaboration with Carl Jacobsen in Copenhagen. In: U.R. Hansson, ed. *Classical Archaeology in the late nineteenth century*. London: De Gruyter.

Moncasoli Tribone, M.L. and M.C. Preacco. 2004. *Luigi Palma di Cesnola. Le gesta di un piemontese dagli scavi di Cipro al Metropolitan Museum of Art*. Turin: Consiglio regionale del Piemonte.

Montelius, O. 1869. *Remains from the Iron Age of Scandinavia*. Stockholm: Ivar Häggström.

Montelius, O. 1883. Några minnen från Sardinien. *Ymer* (1883), pp. 3–5.

Montelius, O. 1885. *Om tidsbestämning inom Bronsåldern med särskildt afseende på Skandinavien*. Stockholm: Kungliga Vitterhets Historie- och Antiqvitets Akademiens handlingar 30.

Montelius, O. 1895–1910. *La civilisation primitive en Italie depuis l'introduction de métaux: illustrée et décrite*. Stockholm: Impr. Royale.

Montelius, O. 1986. *Dating in the Bronze Age: with special reference to Scandinavia*. Stockholm: Kungliga Vitterhets Historie- och Antikvitetsakademien.

Montuori, A. and R.E. Purser. 1995. Deconstructing the lone genius: toward a contextual view of creativity. *Journal of Humanistic Psychology* 35(3), pp. 69–112.

Morávková, N. and V. Řehoř. 2012. *Jindřich Čadík, neprávem zapomenutý Plzeňský historic*. Pilsen: viaCentrum.

Moro Abadía, O. 2009. The history of archaeology as seen through the externalism–internalism debate: historical development and current challenges. *Bulletin of the History of Archaeology* 19(2), pp. 13–26.

Moro Abadía, O. 2013. Thinking about the concept of archive: reflections on the historiography of Altamira. In: O.M. Abadía and C. Huth, eds. *Speaking Materials. Sources for the history of archaeology*, *Complutum* 24(2), pp. 145–52.

Moser, S. 2006. *Wondrous Curiosities: ancient Egypt at the British Museum*. Chicago: University of Chicago Press.

Mulcahy, K. 2017. *Public Culture, Cultural Identity, Cultural Policy: comparative perspectives*. New York: Palgrave Macmillan.

Murley, J. 2012. The impact of Edward Perry Warren on the study and collections of Greek and Roman antiquities in American academia. Unpublished PhD dissertation, Louisville Kentucky.

Murray, T. 2012. Writing histories of archaeology. In: R. Skeates, ed. *The Oxford Handbook of Public Archaeology*. Oxford: Oxford University Press.

Murray, T. 2014. *From Antiquarian to Archaeologist: The history and philosophy of archaeology*. Barnsley: Pen and Sword Books.

Murray, T. and C. Evans, eds. 2008. *Histories of Archaeology. A reader in the history of archaeology*. Oxford: Oxford University Press.

Nagel, T. 1986. *The View from Nowhere*. New York: Oxford University Press.

Nagy, H. 2008. Etruscan votive terracottas from Cerveteri in the Museum of Fine Arts, Boston. A glimpse into the history of the collection. *Etruscan Studies* 11, pp. 101–19.

Naylor, S. 2002. The field, the museum and the lecture hall: the spaces of natural history in Victorian Cornwall. *Transactions of the Institute of British Geographers* 27(4), pp. 494–513.

Naylor, S. 2005. Introduction: historical geographies of science – places, contexts, cartographies. *British Journal for the History of Science* 38(1), pp. 1–12.

New Orleans Academy of Sciences. 1854. *Proceedings*. Vol. 1. New Orleans.

New York Times. 1925. Czechs find temple and beautiful statues dating to 200BC on site of ancient Kyme. *New York Times*, 5 December, p. 21.

Newman, M.E.J. 2001. The structure of scientific collaboration networks. *Proceedings of the National Academy of Sciences of the United States of America* 98(2), pp. 404–9.

Nikić, Lj., G. Žujović and G. Radojčić-Kostić. 2007. *Material for the Bibliographical Dictionary of the members of the Society of Serbian Letters, Serbian Learned Society and Serbian Royal Academy 1841–1947*. Belgrade: Serbian Academy of Sciences and Arts.

Nikolić, D., and J. Vuković. 2008. Od prvih nalaza do metropole kasnog neolita: otkriće Vinče i prva istraživanja/Од првих налаза до метрополе касног неолита:

откриће Винче и прва истраживања. In: D. Nikolić, ed. *Vinča, praistorijska metropola/Винча, праисторијска метропола*. Belgrade: Filozofski fakultet Univerziteta u Beogradu, pp. 39–86.

Nikolić, M.M. 1979. Janko Šafarik (1814–1876), *Godišnjak narodne biblioteke Srbije za 1978*, pp. 11–28.

Nordström, P. 2014. *Arkeologin & livet: ett dubbelporträtt av paret Agda och Oscar Montelius genom deras brevväxling 1870–1907*. Stockholm: Atlantis.

Novák, P. 2006. Počátky československo-tureckých vztahů 1918–1926. Rigorozní práce, Univerzita Karlova.

Novaković, P. 2011. Archaeology in the new countries of Southeastern Europe: a historical perspective. In: L.R. Lozny, ed. *Comparative Archaeologies: a sociological view of the science of the past*. New York: Springer, pp. 339–461.

Novaković, P. 2012. The 'German School' and its influence on the national archaeologies of the Western Balkans. In: B. Migotti, P. Mason, B. Nadbath and T. Mulh, eds. *Scripta in honorem Bojan Đurić. Monografije centra za preventivno arheologijo*. Ljubljana: Zavod za varstvo kulturne dediščine, pp. 151–71.

Olšáková, D. 2007. V krajině za zrcadlem, političtí emigranti v poúnorovém Československu a případ Aymonin. *Časopis Soudobé Dějiny* 14(4), pp. 719–43.

Olšáková, D. 2008. Les exilés Français en Tchécoslovaquie dans les années 1950. Colloque Exil et Dissidence en Europe Centrale, Université Libre de Bruxelles, 13 March.

Olsen, B. 2010. *In Defense of Things: archaeology and the ontology of objects*. New York: AltaMira Press.

Orcutt, K.A. 2006. Personal collecting meets institutional vision: the origins of Harvard's Fogg Museum. *Journal of the History of Collections* 18(2), pp. 267–84.

Örma, S. and K. Sandberg, eds. 2011. *Wolfgang Helbig e la scienza dellantichità del suo tempo*. Rome: Institutum Romanum Finlandiae.

Orser, C.E. 2004. *Historical Archaeology*. 2nd edn. New York: Pearson.

Ort, A. and S. Regourd, eds. 1994. *La rôle de la France dans la création de l'état Tchécoslovaque*. Toulouse: Presse de l'Institut d'études politiques de Toulouse.

Palavestra, A. 2012. Vasić pre Vinče (1900–1908). *Etnoantropološki Problemi* 7(3), pp. 649–79.

Palavestra, A. 2013. Čitanja Miloja M. Vasića u srpskoj arheologiji. *Etnoantropološki problemi* 8(3), pp. 681–715.

Palavestra, A. and S. Babić. 2016. 'False Analogy': transfer of theories and methods in archaeology (the case of Serbia). *European Journal of Archaeology* 19(2), pp. 316–34.

Palavestra, A. and Milosavljević, M. 2015. Miloje M. Vasić i srpska arheologija 1901–1914/Милоје М. Васић и српска археологија 1901–1914. In: M. Ković, ed. *Srbi 1903–1914: Istorija ideja/Срби 1903–1914: Историја идеја*. Belgrade: Clio, pp. 319–27.

Palombi, D. 2006. *Rodolfo Lanciani. L'archeologia a Roma tra ottocento e novecento*. Rome: L'Erma di Bretschneider.

Pasqui, A. 1900. Todi: nuove scoperte nella località di Vasciano. *Notizie Degli Scavi di Antichità*, pp. 251–2.

Patterson, T.C. 1993. *Archaeology: the historical development of civilizations*. Englewood Cliffs, NJ: Prentice Hall.

Patterson, T.C. 1995. *Toward a Social History of Archaeology in the United States*. Fort Worth, Texas: Harcourt Brace College Publishers.

Patterson, T.C. and C.E. Orser, Jr. 2004. Introduction: V. Gordon Childe and the foundations of social archaeology. In: T.C. Patterson and C.E. Orser, Jr, eds. *Foundations of Social Archaeology. Selected writings of V. Gordon Childe*. Oxford: Berg, pp. 1–23.

Pavlowitch, S.K. 1999. *A History of the Balkans 1804–1945*. London: Routledge.

Perrot, G. 1900. Bibliographie. *Revue Archéologique*, pp. 475–84.

Petrie, W.M.F. 1896. *A History of Egypt*, vol. II: *The XVIIth and XVIIIth dynasties*. London, Methuen and Co.

Petrie, W.M.F. and J.E. Quibell. 1896. *Naqada and Ballas, 1895*. London: Bernard Quaritch.

Petrovich, M.B. 1976. *A History of modern Serbia 1804–1918*, vol. I. New York: Harcourt Brace Jovanovich.

Petrović, P. and M. Vasić. 1996. The Roman frontier in Upper Moesia: archaeological investigations in the Iron Gate area – main results. In: P. Petrović, ed. *Roman Lines on the Middle and Lower Danube*. Belgrade: Archaeological Institute, pp. 15–26.

Piggott, S. 1950. *William Stukeley, an eighteenth-century antiquary*. Oxford: Clarendon Press.

Piggott, S. 1958. The excavation of the West Kennet Long Barrow: 1977–6. *Antiquity* 32, pp. 235–42.

Piggott, S. 1962. *The West Kennet Long Barrow: excavations 1955–6*. London: Her Majesty's Stationery Office.

Piggott, S. 1976. *Ruins in a Landscape: essays in antiquarianism*. Edinburgh: Edinburgh University Press.

Platz-Horster, G. 2005. *Lantica maniera: zeichnungen und gemmen des Giovanni Calandrelli in der antikensammlung Berlin*. Berlin and Cologne: S.M.B. Dumont.

Pollak, L. 1994. *Römische Memoiren: Künstler, Kunstliebhaber und Gelehrte 1893–1943*, ed. M. Merkel Guldan. Rome: L'Erma di Bretschneider.

Pollard, J. and A. Reynolds. 2002. *Avebury: The biography of a landscape*. Stroud: Tempus.

Porter, J.L. 2006. What is classical about classical antiquity? In: J.L. Porter, ed. *Classical Pasts: the classical traditions of Greece and Rome*. Princeton, NJ: Princeton University Press, pp. 27–61.

Poulsen, F. 1927. *Das Helbig Museum der Ny Carlsberg Glyptothek. Beschreibung der etruskischen Sammlung*. Copenhagen: Gedruckt bei Neilsen and Lydiche.

Powers, S. 1877. Centennial mission to the Indians of Western Nevada and California. *Annual Report of the Board of Regents of the Smithsonian Institution for 1876*. Washington, DC: US Government Printing Office, pp. 449–460.

Pruitt, T. 2011. Authority and the production of knowledge in archaeology. Unpublished PhD dissertation. University of Cambridge.

Purkhart, M. 2010. Einen neuen Handlungszweig nach der Levante durch Verfertigung der Türkischen Kappen zu veruschen – Österreichische Feze in Konstantinopel. In: R. Agstner and E. Samsinger, eds. *Österreich in Istanbul*. Vienna: Lit Verlag, pp. 259–66.

Putnam, F.W. 1876. Report of the curator. *Tenth Annual Report of the Trustees of the Peabody Museum of American Archaeology and Ethnology*. Cambridge, Mass., pp. 7–12.

Quirke, S. 2010. *Hidden Hands: Egyptian workforces in Petrie excavation archives, 1880–1924*. London: Duckworth.

Raczkowski, W. 2011. The 'German School of Archaeology' in its Central European context: sinful thoughts. In: A. Gramsch and U. Sommer, eds. *A History of Central European Archaeology: theory, methods, and politics*. Budapest: Archaeolingua, pp. 197–214.

Randall-MacIver, D. 1924. *Villanovans and early Etruscans: a study of the early Iron Age in Italy, as it is seen near Bologna, in Etruria and in Latium*. Oxford: Clarendon Press.

Read, M.C. and C. Whittlesey. 1877. Antiquities of Ohio. *Final Report of the Ohio State Board of Centennial Managers to the General Assembly of the State of Ohio*. Columbus: Nevins and Myers, pp. 81–141.

Reid, D.M. 2002. *Whose Pharaohs? Archaeology, museums, and Egyptian identity from Napoleon to World War I*. Berkeley: University of California Press.

Reid, D.M. 2015. *Contesting Antiquity in Egypt: archaeologies, museums, and the struggle for identities from World War I to Nasser*. Cairo: AUC Press.

Reinach, S. 1907a. Necrologie: Adolphe Furtwaengler. *Chronique des Arts*, 19 October, pp. 309–11.

Reinach, S. 1907b. Adolf Furtwaengler. *Revue Archéologique* 3, pp. 326–7.

Reinach, S. 1928. Souvenirs de Furtwaengler. *Revue Archéologique* 5(27), pp. 204–5.

René-Hubert, M. 2010. Des Hellénistes en guerre: Les parcours atypique des membres de l'École française d'Athènes durant la Première Guerre mondiale. *Revue historique des armées*, p. 261.

Rietbergen, P. 2012. *Rome and the World – the World in Rome. The politics of international cuture, 1911–2011*. Dordrecht: Republics of Letters.

Riezler, W. 1965. Adolf Furtwängler zum Gedächtnis. In: A. Greifenhagen, ed. *Adolf Furtwängler: Briefe aus dem Bonner Privatdozentjahr 1879/80 und der Zeit seiner Tätigkeit an den Berliner Museen 1880–1894*. Stuttgart: W. Kolhammer.

Roberts, J. 2005. Towards a social history of archaeology. Unpublished PhD dissertation. University of Wales, College Newport.

Robinson, E. 1889. *Description of twenty-three objects found on the site of Artemisium of Nemi Nemus Dianae during the excavations of Sig. Luigi Boccanera in the spring of 1887 and now in the Museum of Fine Arts in Boston*. Boston.

Roos A.G. Levensbericht F. Cumont, eds. 1950–51. *Jaarboek der Koninklijke Akademie van Wetenschappen*. Amsterdam: Noord-Hollandsche Uitgeversmaatschappij.

Rotenstreich, N. 1986. The proto-ideas and their aftermath. In: R.S. Cohen and T. Schnelle, eds. *Cognition and Fact, materials on Ludwik Fleck*. Boston Studies in the Philosophy of Science 87. Dordrecht: Springer, pp. 161–78.

Rouet, P. 2001. *Approaches to the Study of Attic Vases: Beazley and Pottier*. Oxford: Oxford University Press.

Roversi, L. 1898. *Luigi Palma di Cesnola e il Metropolitan Museum of Art di New York*. New York: Metropolitan Museum of Art.

Rudwick, M. 1985. *The Great Devonian Controversy: The shaping of scientific knowledge among gentlemanly specialists.* Chicago: University of Chicago Press.

Rybczynski, W. and L. Olin. 2007. *Vizcaya. An American villa and its makers.* Philadelphia: University of Pennsylvania Press.

Rydh, H. 1926a. *Grottmänniskornas årtusenden.* Stockholm: Norstedts.

Rydh, H. 1926b. *Kvinnan i Nordens forntid.* Stockholm: Natur och Kultur.

Rydh, H. 1927. *Solskivans land.* Stockholm: Natur och Kultur.

Rydh, H. 1928. *Kring Medelhavets stränder.* Stockholm: Natur och Kultur.

Rydh, H. 1929a. On symbolism in mortuary ceramics. *Bulletin of The Museum of Far Eastern Antiquities* 1, pp. 71–124

Rydh, H. 1929b. *The Land of the Sun-God: Descriptions of ancient and modern Egypt.* London: George Allen and Unwin.

Rydh, H. 1930. *Mor berättar om hur det var förr i världen.* Stockholm: Natur och Kultur.

Rydh, H. 1931. Seasonal fertility rites and the death cult in Scandinavia and China. *Bulletin of The Museum of Far Eastern Antiquities* 3, pp. 69–98.

Rydh, H. 1937. *En vägrödjare genom årtusenden.* Stockholm: Åhlén & söners förlag.

Rydh, H. and B. Schnittger. 1927. *Aranaes: en 1100-Talsborg i Västergötland.* Stockholm: Kungliga Vitterhets-, historie- och antikvitetsakademien.

Rydh, H. and B. Schnittger. 1928. *Där fädrens kummel stå: en utfärdsbok för stockholmstraktens fornlämninga,* vol. 2. Stockholm: Natur och Kultur.

Said, E.W. 1978. *Orientalism.* New York: Pantheon Books.

Salač, A. 1915. Chrámy Egyptských božstev na Delu. *Listy filologické* 42(6), pp. 401–21.

Salač, A. 1920. Momentky z cesty do Řecka: '*Národní listy*'. *Národní listy*, evening edn, 17 August, p. 1.

Salač, A. 1923. Česko-Francouzské vykopávky na Samothrace. *Národní listy*, evening edn, 16 December, p. 9.

Salač, A. 1926a. Obrázky z naší výpravy do Malé Asie. *Národní listy*, evening edn, 5 January, p. 5.

Salač, A. 1926b. Dr A. Salač: z naší archeologické výpravy do malé Asie. I, *Světozor* 26(24), pp. 474–5.

Salač, A. 1926c. Dr A. Salač, z naší archeologické výpravy do malé Asie. II, *Světozor* 26(25), pp. 496–7.

Salač, A. 1927. Inscriptions de Kymé d'Éolide, de Phocée, de Tralles et de quelques autres villes d'Asie Mineure. *Bulletin de Correspondance Hellénique* 51, pp. 374–400.

Salač, A. 1940. Nad rakví Františka Groha. *Listy Filologické* 67(5), pp. 409–11.

Salomonson, J.W. 1969. Review of M.J. Vermaseren and C.C. van Essen 1965. *Mnemosyne* 23, pp. 107–12.

Sangiorgi, G. 1968. *S. Prisca e il suo mitreo.* Rome: Marietti.

Sassatelli, G. 2015. *Archaeologia e preistoria: alle origini della nostra disciplina. Il congresso di Bologna del 1871 e i suoi protagonisti.* Bologna: Bononia University Press.

Savage, R., S. Johnson and G. Phillis. 1780. *The Poetical Works of Richard Savage, with the life of the author.* Edinburgh: Apollo Press.

Sayer, D. 1996. The language of nationality and the nationality of language: Prague 1780–1920. *Past & Present* 153, pp. 164–210.

Schlanger, N. 2006. Introduction. Technological commitments: Marcel Mauss and the study of techniques in the French social sciences. In: N. Schlanger, ed. *Marcel Mauss: techniques, technology and civilisation*. New York and Oxford: Berghahn Books, pp. 1–29.

Schlanger, N. and J. Nordbladh. 2008. *Archives, Ancestors, Practices: Archaeology in the light of its history*. New York: Berghahn Books.

Schmidt, P.R. and T.C. Patterson, eds. 1995. *Making Alternative Histories: The practice of archaeology and history in non-Western settings*. Santa Fe, NM: School of American Research Press.

Schnapp, A. 1996. *The Discovery of the Past: the origins of archaeology*. London: British Museum Press.

Schnapp, A. 2002. Between antiquarians and archaeologists – continuities and ruptures. *Antiquity* 76, pp. 134–40.

Schuchhardt, W-H. 1956. *Adolf Furtwängler*. Freiburger Universitätsreden, N.F. 22. Freiburg: Hans Ferdinand Schulz Verlag.

Secord, J. 2000. *Victorian Sensation: the extraordinary publication, reception, and secret authorship of 'The vestiges of the natural history of creation'*. Chicago: University of Chicago Press.

Semple, S.J. 2003. Burials and political boundaries in the Avebury region, North Wiltshire. *Anglo-Saxon Studies in Archaeology and History* 12, pp. 72–91.

Shapin, S. 1998. Placing the view from nowhere: historical and sociological problems in the location of science. *Transaction of the Institute of British Geographers* 23, pp. 1–8.

Shapin, S. 2010. *Never Pure: historical studies of science as if it was produced by people with bodies, situated in time, space, culture, and society, and struggling for credibility and authority*. Baltimore, Mary.: Johns Hopkins University Press.

Shepherd, N. 2002. The politics of archaeology in Africa. *Annual Review of Anthropology* 31, pp. 189–209.

Shepherd, N. 2003. 'When the hand that holds the trowel is black': disciplinary practices of self-representation and the issue of 'native' labour in archaeology. *Journal of Social Archaeology* 3, pp. 334–52.

Sheppard, K. 2010. Flinders Petrie and eugenics. *Bulletin of the History of Archaeology* 20(1), pp. 16–29.

Sheppard, K. 2013. *The Life of Margaret Alice Murray: A woman's work in archaeology*. Lanham, Mary.: Lexington Books.

Sheppard, K. 2018. *My dear Miss Ransom...: letters between Caroline Ransom Williams and James Henry Breasted, 1898–1935*. Oxford: Archaeopress.

Sieveking, J. 1907. Adolf Furtwängler. *Biographisches Jahrbuch und Deutscher Nekrolog* 12, pp. 188–91.

Sieveking, J. 1909. Adolf Furtwängler. *Biographisches Jahrbuch für die Altertumswissenschaft* 32, pp. 119–31.

Škorić, M. 2010. *Sociologija nauke. Mertonovski i konstruktivistički programi*. Sremski Karlovci: Izdavačka knjižnica Zorana Stojanovića.

Smith, A.C. 1884. *British and Roman Antiquities of North Wiltshire*, vol. 1. Devizes: The Marlborough College Natural History Society.

Smith, M.D. 1996. *A Museum: the history of the cabinet of curiosities of the American Philosophical Society*. Philadelphia, Penn.: American Philosophical Society.

Smith, P.J. 1997. Professor Dorothy A.E. Garrod: 'small, dark, and alive!' *Bulletin of the History of Archaeology* 7(1), pp. 1–2. DOI: http://doi.org/10.5334/bha.07102.

Smith, P.J. 2000. Dorothy Garrod, first woman professor at Cambridge. *Antiquity* 74(283), pp. 131–36.

Smith, P.J. 2009. *'A Splendid Idiosyncrasy': prehistory at Cambridge 1915–1950*. Oxford: Archaeopress.

Smithsonian Institution. 1877. Additions to the collections of the National Museum, in charge of the Smithsonian Institution in 1876. In: *Annual Report of the Board of Regents of the Smithsonian Institution for 1876*. Washington, DC: US Government Printing Office, pp. 84–105.

Sommer, U. 2009. The International Congress of Prehistoric Anthropology and Archaeology and German archaeology. In: M. Babes and M-A. Kaeser, eds. *International Congress of Prehistoric and Protohistoric Sciences 2009. Archaeologists without boundaries: towards a history of international archaeological congresses (1866–2006)*. Oxford: Archaeopress.

Sondheimer, J.H. 1958. *History of the British Federation of University Women, 1907–1957*. London: BFUW.

Sowards, S.W. 2004. *Moderne Geschichte des Balkans. Der Balkan im Zeitalter des Nationalismus*. Seuzach: Books on Demand.

Sox, D. 1991. *Bachelors of Art. Edward Perry Warren & the Lewes House Brotherhood*. London: Fourth Estate.

Squier, E.G. 1850. Aboriginal monuments of the State of New-York. *Smithsonian Contributions to Knowledge*, vol. 2. Washington, DC.

Squier, E.G. and E.H. Davis. 1848. Ancient monuments of the Mississippi Valley. *Smithsonian Contributions to Knowledge*, vol. 1. Washington, DC.

Srejović, D. 1973. Karagač and the problem of the ethnogenesis of the Dardanians. *Balcanica* 4, pp. 39–82.

Srejović, D. 1979. Pokušaj etničkog i teritorijalnog razgraničenja starobalkanskih plemena na osnovu načina sahranjivanja. In: M. Garašanin, ed. *Sahranjivanje kod Ilira*. Belgrade: Balkanološki institut SANU, pp. 79–87.

Státní Úřad Statistický, ed. 1924. *Sčítaní lidu v'republice československé ze dne 15. února 1921*. Prague: Státní Úřad Statistický.

Stephens, S.A. and P. Vasunia, eds. 2010. *Classics and National Cultures*. New York: Oxford University Press.

Stevenson, A. 2019. *Scattered Finds: archaeology, Egyptology and museums*. London: UCL Press.

Stocking, G. 1976. The first American Anthropological Association. *History of Anthropology Newsletter* 3(1), pp. 7–10.

Stolberg, E-M. 2008. Balkan crises (1875–1878, 1908–1913). In: C. Cavanagh Hodge, ed., *Encyclopedia of the Age of Imperialism, 1800–1914*, vol. 1. Westport, Conn.: Greenwood Press, pp. 67–9.

Stout, A. 2008. *Creating Prehistory: druids, ley hunters and archaeologists in pre-war Britain*. Oxford: Blackwell.

Straub. E. 2007. *Die Furtwänglers: Geschichte einer deutschen Familie*. Munich: Siedler.

Stray, C. 1995. Jesse Crum's tour of Greece, Easter 1901. *Classics Ireland* 2, pp. 121–31.

Strenski, I. 1985. What structural mythology owes to Henri Hubert. *Journal of the Behavioral Sciences* 21(4), pp. 354–71.

Strenski, I. 1987. Henri Hubert, racial science and political myth. *Journal of the Behavioral Sciences* 23(4), pp. 353–67.

Strouse, J. 2000. *J. Pierpont Morgan: financier and collector.* New York: Metropolitan Museum of Art.

Studniczka, F. 1907. *Adolf Furtwängler: zum Winckelmannfeste des archäologischen Seminars der Universität Leipzig (neue Jjahrbücher für das klassische Altertum. Gedichte und Deutsche Literatur).* Leipzig: Teubner.

Stukeley, W. 1743. *Abury: A temple of the druids with some others described.* London: printed for the author.

Sugg, R. 2008. Corpse Medicine: mummies, cannibals and vampires. *The Lancet* 371, pp. 2078–9.

Sugg, R. 2011. *Mummies, Cannibals and Vampires: the history of corpse medicine from the Renaissance to the Victorians.* London: Routledge.

Sugg, R. 2013. Medicinal cannibalism in early modern literature and culture. *Literature Compass* 10(11), pp. 825–35.

Sweet, R. 2004. *Antiquaries: the discovery of the past in eighteenth-century Britain.* London: Bloomsbury.

Taylor, C.W. and F. Barron, eds. 1963. *Scientific Creativity: its recognition and development.* New York: Wiley.

Taylor, W. 1948. *A Study of Archaeology.* Menasha, Wis.: American Anthropological Association.

Teichner, F. 2015. Balkanarchäologie – Spiegel der Zeitgeschichte? In: G. von Bülow, ed. *Kontaktzone Balkan: Beiträge des internationalen Kolloquiums die Donau-Balkan-Region als Kontaktzone zwischen Ost-West und Nord-Süd vom 16–18 Mai 2012 in Frankfurt a. M.* Bonn: Habelt, pp. 1–31.

Terrall, M. 2014. *Catching Nature in the Act: Réaumur and the practice of natural history in the eighteenth century.* Chicago: University of Chicago Press.

Thompson, J. 1992. *Sir Gardner Wilkinson and his Circle.* Austin: University of Texas Press.

Thompson, J. 2008. *A History of Egypt; from earliest times to the present.* Cairo: AUC Press.

Thompson, J. 2015. *Wonderful Things: A history of Egyptology*, vol. 1, *From antiquity to 1881.* Cairo: AUC Press.

Thurnam, J. 1860. On the examination of a chambered long barrow at West Kennet, Wiltshire. *Archaeologia* 38, pp. 405–21.

Timotijević, M. 2011. Visuelle Darstellung Serbiens in Werke von Felix Kanitz. In: Đ. Kostić, ed. *Balkanbilder von Felix Kanitz / Слике са Балкана Феликса Каница.* Belgrade: Nationalmuseum, pp. 91–123.

Todorova, M. 2009 [1997]. *Imagining the Balkans.* Oxford: Oxford University Press.

Tomkins, C. 1989. *Merchants and Masterpieces. The story of the Metropolitan Museum of Art.* New York: Henry Holt and Company.

Trenn, T.J. and R.K. Merton. 1981. Biographical sketch. In: R.K. Merton and T.J. Trenn, eds. *Genesis and Development of a Scientific Fact.* Chicago: University of Chicago Press, pp. 149–53.

Trigger, B. 1980. *Gordon Childe: revolutions in archaeology*. London: Thames and Hudson.

Trigger, B. 1989. *A History of Archaeological Thought*. Cambridge: Cambridge University Press.

Tyskiewicz, M. 1895. Notes et souvenirs d'un vieux collectionneur. *Revue Archéologique* 27, pp. 273–85.

Tyskiewicz, M. 1896a. Notes et souvenirs d'un vieux collectionneur. *Revue Archéologique* 28, pp. 6–16.

Tyskiewicz, M. 1896b. Notes et souvenirs d'un vieux collectionneur. *Revue Archéologique* 29, pp. 198–203.

Tyskiewicz, M. 1897a. Notes et souvenirs d'un vieux collectionneur. *Revue Archéologique* 30, pp. 1–7.

Tyskiewicz, M. 1897b. Notes et souvenirs d'un vieux collectionneur. *Revue Archéologique* 31, pp. 166–71.

Ucko, P. and T. Champion. 2003. *The Wisdom of Egypt: changing visions through the ages*. London: UCL Press.

Vasić, M. 1929. Feliks Kanic (Povodom stogodišnjice rođenja)/Феликс Каниц (Поводом стогодишњице рођења). *Srpski književni glasnik* 27, pp. 594–603.

Vasić, M.M. 1955. Prikaz Povodom jedne knjige:/Приказ Поводом једне књиге: V. Gordon Childe, l'aube de la civilisation Europėene. *Starinar* n.s. III–IV, pp. 233–40.

Valenti, C. 2001. L'Ecole Francaise d'Athènes pendant la grande guerre. *Guerres Mondiales et Conflits Contemporains* 204, pp. 5–14.

Valenti, C. 2006. *L'Ecole Française d'Athènes*. Paris: Belin.

Vasić, M.M. 1952. Review of V. Gordon Childe, *The Dawn of European Civilization*. *Starinar* 3–4, pp. 233–40.

Vermaseren, M.J. 1953. *Opgravingen te Rome*. Amsterdam: Stichting IVIO.

Vermaseren, M.J. 1955. Another Mithras temple uncovered: recent discoveries in Rome. *The Illustrated London News*, 8 January, pp. 60–1.

Vermaseren, M.J. 1959. *Mithras de geheimzinnige God*. Amsterdam: Elzevier.

Vermaseren, M.J. 1963. *Mithras, the Secret God*. London: Chatto and Windus.

Vermaseren, M.J. 1975. Nuove indagini nellarea della Basilica di Santa Prisca a Roma. *Mededelingen van het Nederlands Instituut te Rome* 37, pp. 87–96.

Vermaseren, M.J. and C.C. van Essen. 1955–6. The Aventine *mithraeum* adjoining the church of St Prisca. A brief survey of the Dutch excavations on the Aventine. *Antiquity and Survival* 1, pp. 3–36.

Vermaseren, M.J. and C.C. van Essen. 1965. *The Excavations in the Mithraeum of the Church of Santa Prisca in Rome*. Leiden: Brill.

Vermeule, C.C. 1981. *Greek and Roman Sculpture in America. Masterpieces in public collections in the United States and Canada*. Berkeley: University of California Press.

Vitali, D. 1984. La scoperta di Villanova e il Conte Giovanni Gozzadini. *Dalla stanza delle antichità al museo civico: storia della formazione del Museo Archeologico di Bologna*: catalogue of the exhibition, pp. 223–41.

Viviers, D. 1996. Un enjeu de politique scientifique: la section étrangère de l'École Française d'Athènes. *Bulletin de Correspondance Hellenique* 120(1), pp. 173–90.

Von Hauer, F. 1882. Zur Erinnerung an Dr Ami Boué. *Jahrbuch der k.k. Geologischen Reichsanstalt* 32, pp. 1–6.

Wagemakers, B. 2015. Fallen into oblivion: the 1956 Greek-Dutch Expedition to Archanes on Crete. *Bulletin of the History of Archaeology* 25(1), DOI: http://doi.org/10.5334/bha.251

Waingart, S.B. 2015. Finding the history and philosophy of science. *Erkenntnis* 80(1), pp. 201–13.

Wehgartner, I. 2001. Spurensuche: Frauen in der Klassischen Archäologie vor dem Ersten Weltkrieg. In: J.K. Koch and E-H. Mertens, eds. *Eine Dame zwischen 500 Herren: Johanna Mestorf, Werk und Wirkung*. Münster: Waxmann, pp. 267–79.

Weissmann, G. 2002. In quest of Fleck, science from the Holocaust. In: G. Weissmann, ed. *Darwin's Audubon: science and the Liberal imagination*. Cambridge: Perseus Publishing, pp. 107–14.

Whitehill, W.M. 1970. *Museum of Fine Arts Boston. A centennial history*, vol. 1. Cambridge Mass.: Belknap Press.

Whitling, F.W. 2018. *Western Ways: Foreign schools in Rome and Athens*. Berlin and New York: De Gruyter.

Whittlesey, C. 1850. Descriptions of ancient works in Ohio. *Smithsonian Contributions to Knowledge*, vol. 3. Washington, DC.

Wilcox, D. and C.M. Hinsley, Jr. 2003. Appendix: Analysis of Registered Members of the International Congress of Anthropology, World's Columbian Exposition, 1893. In: D. Wilcox and C.M. Hinsley, Jr. *The 1893 World's Fair and the Coalescence of American Anthropology*. Chicago: University of Chicago Press, pp. 110–19.

Willers, H. 1901. Review of Furtwängler 1900. *Berliner Philologische Wochenschrift* 21, pp. 1103–10, 1135–41, 1168–74.

Willey, G. and J. Sabloff. *A History of American Archaeology*. San Francisco: W.H. Freeman.

Williams, T.J.T. 2015. Landscape and warfare in Anglo-Saxon England and the Viking campaign of 1006. *Early Medieval Europe* 23(3), pp. 329–59.

Withers, C. and D. Livingstone. 2011. Thinking geographically about nineteenth-century science. In: D. Livingstone and C. Withers, eds. *Geographies of Nineteenth-century Science*. Chicago: University of Chicago Press, pp. 1–19.

Wolters, P. 1910. Adolf Furtwängler: Gedächtnisrede gehalten in der Festsitzung der kgl. Bayer Akad. der Wissenschaften am 20. Nov. 1909. *Süddeutsche Monatshefte* 7, pp. 90–105.

Wright, A. 1977. Review: *The Religion of the Chinese People* by Marcel Granet and Maurice Freedman. *Pacific Affairs* 49(4), pp. 696–7.

Wünsche, R. 2007. Adolf Furtwängler (1853–1907): Archäologe, Direktor der Glyptothek, Ausgräber von Ägina. *Oberbayerisches Archiv* 132, pp. 299–337.

Zahra, T. 2004. Reclaiming children for the nation: Germanization, national ascription, and democracy in the Bohemian lands, 1900–1945. *Central European History* 37(4), pp. 501–43.

Zazoff, P. 1983. *Gemmensammler und Gemmenforscher: von einen noblen Passion zur Wissenschaft*. Munich: Beck.

Index